Introductory Tiling Theory for Computer Graphics

Synthesis Lectures on Computer Graphics and Animation

Editor
Brian A. Barsky, *University of California, Berkeley*

Introductory Tiling Theory for Computer Graphics
Craig S. Kaplan
2009

Practical Global Illumination with Irradiance Caching
Jaroslav Krivanek, Pascal Gautron
2009

Wang Tiles in Computer Graphics
Ares Lagae
2009

Virtual Crowds: Methods, Simulation, and Control
Nuria Pelechano, Jan M. Allbeck, Norman I. Badler
2008

Interactive Shape Design
Marie-Paule Cani, Takeo Igarashi, Geoff Wyvill
2008

Real-Time Massive Model Rendering
Sung-eui Yoon, Enrico Gobbetti, David Kasik, Dinesh Manocha
2008

High Dynamic Range Video
Karol Myszkowski, Rafal Mantiuk, Grzegorz Krawczyk
2008

GPU-Based Techniques for Global Illumination Effects
László Szirmay-Kalos, László Szécsi, Mateu Sbert
2008

High Dynamic Range Image Reconstruction
Asla M. Sá, Paulo Cezar Carvalho, Luiz Velho
2008

High Fidelity Haptic Rendering
Miguel A. Otaduy, Ming C. Lin
2006

A Blossoming Development of Splines
Stephen Mann
2006

Introductory Tiling Theory for Computer Graphics
Craig S. Kaplan

ISBN: 978-3-031-79542-8   paperback
ISBN: 978-3-031-79543-5   ebook

DOI 10.1007/978-3-031-79543-5

A Publication in the Springer series
SYNTHESIS LECTURES ON COMPUTER GRAPHICS AND ANIMATION

Lecture #11
Series Editor: Brian A. Barsky, *University of California, Berkeley*

Series ISSN
Synthesis Lectures on Computer Graphics and Animation
Print 1933-8996   Electronic 1933-9003

Introductory Tiling Theory for Computer Graphics

Craig S. Kaplan
University of Waterloo

SYNTHESIS LECTURES ON COMPUTER GRAPHICS AND ANIMATION #11

ABSTRACT

Tiling theory is an elegant branch of mathematics that has applications in several areas of computer science. The most immediate application area is graphics, where tiling theory has been used in the contexts of texture generation, sampling theory, remeshing, and of course the generation of decorative patterns. The combination of a solid theoretical base (complete with tantalizing open problems), practical algorithmic techniques, and exciting applications make tiling theory a worthwhile area of study for practitioners and students in computer science.

This synthesis lecture introduces the mathematical and algorithmic foundations of tiling theory to a computer graphics audience. The goal is primarily to introduce concepts and terminology, clear up common misconceptions, and state and apply important results. The book also describes some of the algorithms and data structures that allow several aspects of tiling theory to be used in practice.

KEYWORDS

symmetry, patterns, tilings, tessellations, wallpaper groups, isohedral tilings, substitution tilings, aperiodic tilings, penrose tilings

Contents

Preface

The text in this book has gone through many incarnations over a period of approximately ten years. I began studying symmetry theory and tiling theory around 1998. In the years that followed, I developed an algorithm for Escherization, which was published in the proceedings of SIGGRAPH 2000 [33]. That paper contained my earliest attempt to explain the mathematical structure and implementation details of isohedral tilings. My 2002 doctoral dissertation [30] incorporated many more details about symmetry and tilings, both in the context of Escherization and as more general background material.

In 2006, I taught a graduate course at the University of Waterloo on computer graphics, geometry, and ornamental design (see http://www.cgl.uwaterloo.ca/~csk/cs798/winter2006/). A large portion of that course borrowed on the ideas that appeared in my dissertation. Although I did not explicitly use my thesis to create lecture notes, I referred to it frequently. Then, in 2008, I collaborated with Ares Lagae, Chi-Wing Fu, Victor Ostromoukhov, Johannes Kopf and Oliver Deussen on a SIGGRAPH course entitled "Tile-based methods for interactive applications" [38]. I built a set of notes for that course based on a distillation of the relevant sections of my dissertation. This book follows directly from those course notes, after a healthy amount of reorganization, additions, and general improvements.

This book aims to provide an accessible introduction to tiling theory, with an emphasis on computational techniques that could be useful to researchers and practitioners in computer graphics. It would be suitable as part of an advanced undergraduate or graduate course, particularly one that straddles the line between mathematics, art, and computer science. At times I make reference to advanced mathematical concepts: there is a smattering of analysis, topology, and group theory to be found in the more mathematical parts of the book. These concepts are included primarily to avoid leaving holes in the exposition. For the most part, they can be glossed over without harming the reader's ability to understand later material. On the other hand, I assume that the reader is familiar with the skills and concepts that make up an undergraduate computer science degree, including the mathematical foundations of computer graphics.

By far the most important inspiration for the presentation in this book is *Tilings and Patterns* by Grünbaum and Shephard [24]. That monumental book is the definitive work on tiling theory, and continues to be one of my all-time favourite mathematical texts. Readers will find it cited liberally throughout this book. Unfortunately, *Tilings and Patterns* has been out of print for years, and so an introduction to tiling theory is becoming increasingly difficult to come by. I hope that this book can in part help to reverse that unfortunate situation.

Motivated readers should also consult the new book *The Symmetries of Things* by Conway et al. [6]. They recast symmetry theory and aspects of tiling theory in the language of orbifolds. Orbifolds offer a powerful, intuitive framework for thinking about these topics, and an accessible introduction to this topic is most welcome. Large parts of this book, particularly the chapters about symmetry and isohedral tilings, might profitably be re-expressed in terms of orbifolds, an exercise I leave for the future.

I have made a conscious decision to restrict the focus of this book to topics of potential interest to a wide audience in computer graphics. As a result, some topics, such as tilings in non-Euclidean spaces, 3D space fillers, and combinatorial tiling theory using Delaney symbols, receive no more than passing mention. Readers who are inspired by the tiling theory in this book will be rewarded by looking further into those topics, perhaps by starting with the more advanced chapters of the aforementioned book *The Symmetries of things*.

The exposition in this book has benefitted from the advice and feedback of a large number of people. David Salesin helped me find my voice as a writer, and many of his suggestions have become laws that I obey to this day. The students in my 2006 graduate course endured my attempt to explain many of these concepts; in particular, Paul Church went on to complete a Master's degree in tiling theory under my supervision, and clarified several important concepts. Thanks especially to Oliver Deussen and Chaim Goodman-Strauss, who read a draft of this book and contributed many valuable ideas.

Craig S. Kaplan
University of Waterloo

June 2009

CHAPTER 1

Introduction

Tiling theory, the study of shapes that cover the plane with no gaps or overlaps, is an elegant and aesthetic branch of mathematics. Tilings themselves are an ancient art form; countless historical and contemporary examples motivate us to explain mathematically what was devised by intuition alone. The mathematical theory that results is a compelling blend of geometry, combinatorics, and group theory, with occasional forays into analysis, topology and graph theory. The area is full of unsolved problems that are simple to state but complex and mysterious when considered carefully [7, 31].

Elements of tiling theory have already found exciting applications in computer graphics, from art and ornamental design to sampling and texture synthesis. In computer graphics, we might be able to benefit from many aspects of tiling theory, either by borrowing its theorems to solve problems, or by developing algorithms for generating novel attractive tilings. My personal bias is towards the decorative applications of tilings. My goal, then, is to share the techniques that could be used to produce attractive two-dimensional graphics, both for direct use in illustration or to drive computer-aided manufacturing technology.

The goal of this book is to present the basics of tiling theory in an accessible way, including additional details of interest to those seeking to use tilings in computer graphics applications. In some places, I have taken license to include more involved mathematical details where they are especially worthwhile or to clear up common confusions. For the most part, however, emphasis is given to those aspects of tiling theory that lend themselves readily to software implementation. Thus many deep and fascinating mathematical topics are omitted. For such topics, motivated readers should still consult the definitive reference by Grünbaum and Shephard [24].

1.1 ORGANIZATION

The rest of this book is organized into six chapters.

- **Chapter 2: Tiling basics** covers the elementary properties shared by all tilings, and defines many of the terms that will be used throughout the book.

- **Chapter 3: Symmetry** offers an introduction to symmetry theory in the plane, presents algorithms for rendering symmetric drawings, and discusses the special case where the object whose symmetries are being studied is a tiling. Symmetry theory will be of interest to anyone studying tiling theory. However, only a small part of this chapter is necessary for understanding the rest of the book. It is important to know what symmetry is and the fact that the wallpaper patterns are periodic.

- **Chapter 4: Tilings by polygons** is about several special classes of polygonal tilings that have a long history in art and mathematics, are important for computer graphics, and help support the subsequent development of the isohedral tilings.

- **Chapter 5: Isohedral tilings**, the central chapter of the book, is devoted to an extended presentation of the theory of isohedral tilings. These tilings are especially useful in computer graphics, particularly in decorative applications. I explain how isohedral tilings may be given symbolic descriptions and how those descriptions can be used to formulate software for representing and manipulating tilings.

- **Chapter 6: Nonperiodic and aperiodic tilings** introduces some well-known examples of tilings that do not have periodic symmetry. These tilings usually exhibit a tantalizing sense of order that makes them useful in artistic applications of computer graphics. They are also gaining momentum as a tool for advanced graphics topics such as sampling and texture synthesis.

- **Chapter 7: Survey** concludes with a brief survey of some of the research in computer graphics that has made use of tiling theory.

Chapters 2 through 6 include exercises. These exercises come in a mix of flavours. Some are like puzzles, in that they challenge the reader to exhibit a tiling with a certain property or that acts as a counterexample to a hypothesis. Some are mathematical exercises of varying degrees of difficulty. Some are programming problems that ask for an implementation of an aspect of the exposition in that chapter. Finally, some problems are really suggestions for research projects that could lead to new results in computer graphics or tiling theory. Generally, I use an asterisk to indicate questions for which I do not know the answer.

Not all exercises are geared towards computer graphics—some simply help deepen one's intuition for the properties of tilings. Also, the exercises occasionally rely on concepts that are not explained in this book. When that happens, I provide appropriate references.

CHAPTER 2

Tiling Basics

The most natural property associated with a tiling of the plane is that it should consist of shapes that cover the plane without any overlap. We may provisionally formalize these notions by stating that a set S of shapes *covers* the plane if the union of all shapes in S is the entire plane, and that an *overlap* is a non-empty intersection between two tiles (in which case S has no overlaps if it consists of pairwise disjoint sets). Under this definition, the tilings of the plane are precisely the set-theoretic partitions of the plane as a set of points. This definition is so universally inclusive that we have gained nothing—it is nearly impossible to speak meaningfully of the topological, combinatorial, geometric, or computational properties of tilings. One goal of this chapter is then to modify and refine this definition of tilings into one that is sufficiently constrained that tilings become distinct, worthwhile mathematical objects. I will deliberately add more constraints than are strictly necessary mathematically, in order to arrive at a definition suitable for the kinds of tilings that we encounter in computer graphics. After formulating a practical definition, I explore some of the basic features of tilings that will be useful throughout this book.

2.1 DEFINING TILINGS

Let us begin by restricting the universe of shapes we are willing to allow as tiles. The subsets of the plane that may be thought of intuitively as "shapes" are mathematically simple. We immediately eliminate unbounded or degenerate shapes. We rule out shapes made from multiple disconnected pieces, or from pieces connected only by points or lines; we might just as well treat the individual pieces as separate tiles. We also do not want to consider shapes with holes; if one shape completely encloses another, the inner shape can be regarded as a drawing or "marking" on the outer one.

We can capture these constraints by requiring that every tile be topologically equivalent to a closed unit disc (i.e., that there be a continuous deformation of the plane that maps the closed unit disc to the given shape without cutting or gluing). Each tile is then a bounded subset of the plane that contains its boundary and encloses a finite area. Note that the restriction of tiles to closed topological discs guarantees that every tiling contains precisely a countable infinity of tiles, which we may henceforth refer to as $\{T_1, T_2, \ldots\}$.

The fact that our tiles are closed presents an immediate problem in the initial definition of tilings offered at the opening of this chapter. Just as the real line cannot be covered with disjoint closed intervals, disjoint closed topological discs cannot cover the plane without fighting for control of their boundaries. We will, therefore, relax the definition of overlapping, and permit tiles to have a non-empty intersection if that intersection is confined to the tiles' boundaries. Only when the interiors of tiles intersect is an overlap considered to have occurred.

A further restriction we wish to make is on the sizes of tiles. Here we adopt a loose notion of size sufficient for this purpose. Let T be a tile, topologically equivalent to a disc as discussed above. Because T is bounded, there exists a real number $U_T > 0$ such that T is completely contained in a closed disc of radius U_T. Because T encloses a finite area, there exists a second real number $u_T > 0$ such that T completely encloses a closed disc of radius u_T.

Even if every individual tile is required to be a topological disc, tiles may still grow or shrink without bound "at infinity". Figure 2.1 shows a simple tiling in which both of these problems occur. From

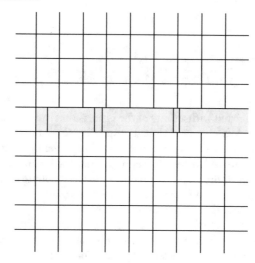

Figure 2.1: An example of a tiling that is not uniformly bounded. All non-shaded tiles are unit squares. The widths of the shaded tiles form the sequence $\{1/2, 2, 1/3, 3, 1/4, 4 \ldots\}$. Although any bounded subset of this tiling is well behaved, in the limit tiles can always be found that are smaller or larger than any bounds.

the point of view of computer graphics, this possibility may not seem like a major concern since we can never draw more than a finite part of any tiling in practice. However, many aspects of tiling theory require that the tilings being studied behave well *everywhere*, not just where we can see them.

   We avoid undesirable behaviour at infinity by requiring not just that individual tiles be bounded, but that all tiles be *uniformly bounded*: there exist real numbers $U > 0$ and $u > 0$, depending only on the tiling, such that *all* tiles enclose a disc of radius u and are enclosed by a disc of radius U. (Grünbaum and Shephard refer to U and u as the *circumparameter* and *inparameter* of the tiling, respectively [24, Section 3.2]).

   Combining the foregoing observations, we arrive at the following definition.

Definition 2.1. Tiling.   A *tiling* is a countable collection $\mathcal{T}$ of *tiles* $\{T_1, T_2, \ldots\}$, such that:
   1. Every tile is a closed topological disk.
   2. Every point in the plane is contained in at least one tile;
   3. The interiors of the tiles are pairwise disjoint; and
   4. The tiles are uniformly bounded.

   As Grünbaum and Shephard point out [24, Section 1.1], it is possible to consider arrangements of shapes that are required to satisfy Condition 2 but not 3, or Condition 3 but not 2, leading to definitions of *coverings* and *packings*, respectively. A tiling is then easily seen as an arrangement of shapes that is simultaneously a packing and a covering. Packings and coverings are of great interest in both mathematics and computer science (for example, there is a deep connection between sphere packings and coding theory), but they are beyond the scope of this book.

2.2   ANATOMY OF A TILING

The *frontier* of a tiling T is the union of the boundaries of all the tiles in T. The frontier will naturally be decomposable into a collection of *tiling vertices*, points that lie on the boundaries of three or more tiles, and *tiling edges*, curves (excluding their endpoints) that begin and end at tiling vertices and belong to exactly two tiles. The boundary of each tile can then be seen as an alternating sequence of tiling vertices and tiling edges. Sometimes, we disregard the incidental shapes of the tiling edges and speak of a tile's *tiling polygon*, the (possibly self-intersecting) polygon connecting the tiling vertices belonging to a single tile.

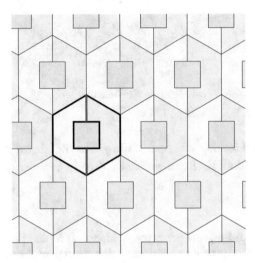

Figure 2.2: An example of a tiling in which some pairs of tiles intersect in disconnected curves. The intersection between the two tiles outlined in bold consists of two disjoint line segments. This tiling is not normal according to the definition given in Section 2.2.

In general, when the intersection of two tiles is non-empty, it may consist of a collection of disconnected curves and points. A simple example, in which some pairs of tiles intersect in two disconnected curves, is illustrated in Figure 2.2. Occasionally, it is useful to restrict our attention to the case where the intersection of two tiles is either empty, a single point, or a single curve connecting two tiling vertices. We can then speak meaningfully of *the* tiling edge shared by two tiles. This restriction also avoids tilings that are problematic topologically, such as that of Figure 2.2, in which there may be multiple tiling edges connecting two tiling vertices. Following Grünbaum and Shephard's terminology we will refer to tilings with this additional restriction as *normal tilings*. The features of normal tilings are illustrated in Figure 2.3.

Let V and E denote the tiling vertices and tiling edges of a normal tiling T. We can then speak of a binary incidence relation over pairs of elements in $T \cup V \cup E$: two elements are related if they have a non-empty intersection (assuming vertices are suitably interpreted as singleton sets). We refer to the set $T \cup V \cup E$ together with this incidence relation as the *combinatorial structure* of T. The combinatorial structure can be thought of as an infinite graph the records all adjacencies between tiling features. Furthermore, T is *combinatorially equivalent* to a second tiling T' with vertices V' and edges E' if there is a bijection between the combinatorial structures of T and T' that maps vertices, edges

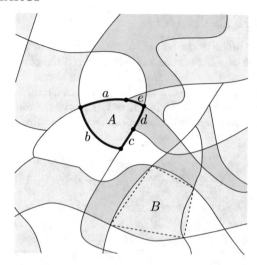

Figure 2.3: Basic topological features of a normal tiling. The tile labeled A is outlined in bold and its tiling vertices are marked with dots. Each of the tiling edges on its boundary is also labeled. Tile B's tiling polygon is shown with dashed lines.

and tiles to vertices, edges and tiles, and preserves incidence. It is also possible to speak of *topological equivalence* between tilings, based purely on continuous deformation of the plane; however, in the case of normal tilings this is unnecessary, as the notions of combinatorial and topological equivalence coincide [24, Section 4.1].

2.3   PATCHES

In any practical application of tiling theory, we will necessarily operate on finite collections of tiles. A *patch* of tiles is a finite set of tiles whose union is topologically equivalent to a disc. That is, a patch is a contiguous block of tiles with no internal holes. (In fact, we ought to be careful here. We may wish to speak of "patches" of shapes without knowing in advance whether they tile the whole plane. For this purpose, a suitable definition may be extracted from the definition of a tiling.)

Of course, no drawing of a tiling shows more than a finite patch, and many show a ring of partial tiles when the drawing is restricted to, say, a rectangular region. We make the implicit assumption that the context will provide the means of understanding how a patch may be extended to cover the plane. When the tiling is highly structured, as in the periodic tilings to be discussed in Section 3.4, the means of extension is immediate; in a chaotic tiling such as that of Figure 2.3, it is irrelevant since only the behaviour over the finite patch is of any interest. Patches are useful in proofs in tiling theory where they can serve as "partial solutions" on the way to tiling the whole plane.

2.4   TILINGS WITH CONGRUENT TILES

In many of the tilings we see every day on walls and streets, the tiles all have (approximately) the same shape. If every tile in a tiling is congruent to some shape T (i.e., there is a rigid motion of the plane, possibly including a reflection, which makes the tile coincide with T), we say that the tiling is *monohedral*, and that T is the *prototile* of the tiling. More generally, a *k-hedral* tiling is one in which every tile is congruent to one of k different prototiles. We also use the terms *dihedral*, *trihedral* and *multihedral* for the cases $k = 2$, $k = 3$ and $k > 1$, respectively. Note that a k-hedral tiling must be uniformly bounded because a suitable circumparameter and inparameter for the tiling can be computed directly from the finite prototile set.

If $\mathcal{P}$ is a set of prototiles, any tiling that can be formed exclusively from congruent copies of members of $\mathcal{P}$ is said to be *admitted* by $\mathcal{P}$. Similarly, a set of prototiles might also be said to admit a given patch.

Note that a tiling admitted by a set of prototiles need not use all of them. Thus k-hedrality must be seen as a property of a tiling, and not of a prototile set. No analogous definition exists for a set of k prototiles; we might ask that the prototiles admit at least one k-hedral tiling, or perhaps that they admit *only* k-hedral tilings. A recurring theme (and frequent source of confusion) in tiling theory is the distinction between a fundamental property of a set of shapes, and an incidental property of one or more tilings that they admit. Typically, a property attached to a set of prototiles is much stronger than the analogous property on a tiling because the former must hold for *all* tilings that the prototiles admit. We will encounter this theme several more times in this book.

Given a finite set of shapes, we might wonder whether they are in fact a prototile set—that is, do the given shapes admit any tilings at all? In full generality, this problem is known to be formally undecidable (see Section 6.2), and so we must speculate that it could be arbitrarily difficult to prove or disprove the fact for any given set of shapes. However, we do have one useful tool at our disposal [24, Section 3.8]:

Theorem 2.2. The Extension Theorem.   *Let $\mathcal{P}$ be a finite set of shapes, each a closed topological disk. If, for any $r > 0$, there exists a patch of tiles from $\mathcal{P}$ that contains a disk of radius r, then $\mathcal{P}$ admits a tiling of the plane.*

The Extension Theorem offers us a kind of generic limiting process: if we can construct arbitrarily large finite patches of tiles, then we can go off to infinity in all directions. This fact is true even if none of the individual patches can be extended to tile the plane, or if the patches are not nested within each other.

Naturally, restricted classes of prototiles can sometimes be shown to tile without the full strength of the Extension Theorem. For example, if pairs of prototiles may be placed next to one another in only finitely many distinct ways (a property known as *finite local complexity*), then a weaker form of the Extension Theorem, based on König's Infinity Lemma, can be used to establish tileability [24, Section 11.2]. Any one prototile set might also come equipped with a specialized argument that yields the tilings it admits.

The Extension Theorem can take the place of many specialized arguments. For example, it can almost always be invoked to justify the construction of substitution tilings (Section 6.1). Of course, the theorem is more important in establishing the existence of a tiling in a mathematical setting. In computer graphics applications, the ability to produce patches of any desired size is sufficient.

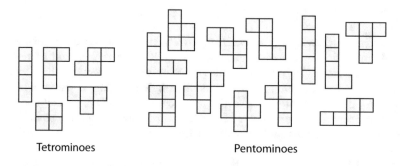

Tetrominoes   Pentominoes

Figure 2.4: The five tetrominoes and 12 pentominoes.

EXERCISES

1. Show that the following shapes tile the plane:

    (a) Any triangle.

    (b) Any (non self-intersecting) quadrilateral.

    (c) Any hexagon consisting of three pairs of opposite, parallel edges.

2. Let S be a family of subsets of the plane, all closed topological discs. Suppose further that for any distinct S_1 and S_2 in S, the intersection of the interiors of S_1 and S_2 is empty. Prove that S must contain a countable number of shapes.

3. Show that for every integer $k \geq 3$ there exists a monohedral tiling by a polygon with k sides. This problem can most easily be solved by finding a small number of infinite families of tilings.

4. A *polyomino* is a connected union of squares taken from an infinite grid of squares. For example. the *tetrominoes*, polyominoes made up of four squares, are the familiar pieces from the game *Tetris*. The tetrominoes and pentominoes are shown in Figure 2.4.

    (a) Exhibit a tiling for each of the tetrominoes.

    (b) Exhibit a tiling for each of the pentominoes.

    (c) A prototile is *monomorphic* if there exists exactly one tiling from that prototile (that is, all tilings admitted by the prototile are congruent to each other). Which of the pentominoes are monomorphic?

    (d) What is the smallest n for which there exists an n-omino that does not tile the plane? Ignore any n-ominoes that have internal holes.

    *(e) Let $P(n)$ represent the number of distinct n-ominoes, and let $T(n)$ represent the number of n-ominoes that are prototiles of monohedral tilings of the plane. Investigate the behaviour of $T(n)$ as n grows. Is there some sense in which "most" n-ominoes do not tile the plane? (This problem probably requires significant new research; for example, the state of the art offers only a fairly coarse estimate for $P(n)$.)

5. (a) Draw (a patch of) a tiling containing a tile with a self-intersecting tiling polygon.

*(b) Can there exist a tiling in which every tile has a self-intersecting tiling polygon?

6. Give an example of a tiling in which no two tiles are congruent. The tiling will have to be described via a mathematical rule that gives the shapes of all tiles in the plane.

7. Let $\{T_1, \ldots, T_n\}$ be a prototile set for a tiling of the plane in which T_n occurs only finitely many times. Prove that the smaller set $\{T_1, \ldots, T_{n-1}\}$ also tiles the plane.

8. Suppose that you wish to write a library in an object oriented language for manipulating and rendering tilings. As a basis, you decide to create a class `Tiling` to act as an abstract base class for all tilings.

   What information and behaviours might be stored in this class? That is, what methods could you put in `Tiling` that would make sense for *all* tilings?

CHAPTER 3

Symmetry

Symmetry is a pervasive concept in modern mathematics, an elegant and powerful tool that can be applied in a broad range of situations. It should not be surprising that there is a strong connection between symmetry and tilings—tilings of the plane typically feature some degree of repetition, and symmetry is a means of measuring that repetition. Planar symmetry groups have served as a powerful tool in understanding and classifying designs belonging to many of the world's ornamental design traditions [51].

In this chapter, I offer an overview of symmetry theory in the plane. I make an effort to fill in some of the underlying intuition for symmetry, and include additional details about how the theory applies in the special case of tilings.

As it happens, the rest of this book does not rely heavily on a direct understanding of symmetry theory. For example, while the isohedral tilings discussed in Chapter 5 all belong to the 17 wallpaper groups, their symmetries emerge as a by-product of the interactions between a tile and its neighbours, and do not require direct attention. From an algorithmic point of view, it will suffice to know simply that every isohedral tiling is periodic. Nevertheless, symmetry theory is sufficiently useful, important, and germane that it seems worthwhile to spend some time on the subject here. The subject can be explored in greater detail in many excellent resources, both in print and online. Interested readers can consult the introductory texts by Farmer [15] and Weyl [52], references on symmetry in art and ornament [48, 51], or the new treatment by Conway et al. [6].

3.1   THE SET OF SYMMETRIES

The most common informal definition of symmetry refers to a shape that is balanced on either side of a central line. For example, a capital letter **A** is symmetric because a vertical mirror may be placed along its midline, and the entire letter reconstructed from either half. An alternate way to think about the action of this mirror is that it exchanges the left and right halves of letter (assuming it is silvered on both sides). More generally still, the mirror may be thought of as a transformation of the entire plane which, when applied to the **A**, leaves it unchanged. We accept this mirror reflection as a legitimate source of symmetry, because while the mirror image of an object may be in a different location and may be oppositely oriented (e.g., the mirror image of a left hand is a right hand), we do not think of the mirror as changing an object's shape.

The preceding discussion suggests a general notion of symmetry for figures in the plane. First, we must define the set $\mathcal{M}$ of all transformations of the plane that do not distort shape. Then, given some figure S, we can identify the subset of $\mathcal{M}$ consisting of transformations T that leave S unchanged, that is $\{T \in \mathcal{M} | T(S) = S\}$.

Having decided that all mirror reflections in the plane should be considered as shape-preserving transformations, we can use reflections to generate $\mathcal{M}$. Observe that if T_1 and T_2 are both transformations that preserve shape, then it is reasonable to expect that their composition $T_2 \circ T_1$ does as well. We can, therefore, let $\mathcal{M}$ consist of the reflections together with all possible compositions of reflections (that is, the closure of the set of reflections under composition of functions). With a bit of algebra or geometry, it can be shown that every transformation in this set will belong to one of five classes:

1. **The identity transformation**, which leaves every point where it is;

2. **Reflections** across lines;

3. **Rotations** of the plane around a point by a specified angle;

4. **Translations**, which displace every point by a specified vector; and

5. **Glide reflections**, each of which consists of a reflection across a line followed by a non-zero translation along that line.

No new classes of transformations will be introduced through further composition from those described above. And because every reflection has itself as a well-defined inverse, $\mathcal{M}$ is automatically closed under inverses as well.

Let $d(p, q)$ be the usual measure of Euclidean distance between points p and q in the plane. A transformation T of the plane is called an *isometry* if $d(T(p), T(q)) = d(p, q)$ for all p and q; that is, if T does not distort distances. It is not hard to see that every reflection is an isometry, from which it follows that $\mathcal{M}$ consists entirely of isometries (because the composition of two isometries must yield an isometry). Less obvious is the fact that every isometry in the plane is a composition of reflections, and hence belongs to $\mathcal{M}$ (see Exercise 3). Thus $\mathcal{M}$ is precisely the set of isometries, and we will adopt this set as the shape-preserving transformations from which symmetries may be found. Because they preserve shape, elements of $\mathcal{M}$ are also sometimes referred to as *rigid motions* or just *motions*. Furthermore, the isometries provide a concrete definition of congruence, a term that was used informally in Section 2.4: two shapes are *congruent* if one can be brought into correspondence with the other via an isometry.

3.2   SYMMETRY GROUPS

Let S be any shape (i.e., any subset of the plane). Define the set $\Sigma(S)$ to be the symmetries of S, that is, the isometries of the plane that map S to itself. For many shapes, this set will consist only of the identity isometry. We refer to S as *symmetric* if $\Sigma(S)$ is non-trivial.

If σ_1 and σ_2 are two symmetries of a shape S, then their composition $\sigma_2 \circ \sigma_1$ must be a symmetry as well, along with their inverses σ_1^{-1} and σ_2^{-1}. From this observation (and the associativity of function composition) we can conclude that for any S, $\Sigma(S)$ forms a group, which we refer to as the *symmetry group* of S. Note that the set $\mathcal{M}$ of all isometries forms a group as well, of which the symmetry group of any shape is a subgroup.

Let G be a symmetry group and p any point in the plane. We define the *orbit* of p to be $\{\sigma(p) | \sigma \in G\}$, the set of all locations to which p is transported by transformations in the group. All symmetrically related points will belong to the same orbit. For the purpose of studying patterns and tilings, we are particularly interested in symmetry groups for which the orbits are not too crowded. We refer to a set S of points in the plane as *discrete* if there is a real number $r > 0$ such that no two points in S are closer than distance r from one another (in the language of analysis, we would say that S contains no limit points). A *discrete symmetry group* is then one for which the orbit of every point is a discrete set. A discrete symmetry group is characterized by a minimum angle for all rotations, and a minimum distance by which points are displaced in all translations and glide reflections.

Just as we develop a notion of congruence to identify equivalent shapes, it would be helpful to have a way of saying that two symmetry groups are "essentially the same". It is tempting to ask simply that the groups be isomorphic, but there is no *a priori* reason to assume that this condition is sufficient—there may (and do) exist very different sets of symmetries that happen to have the same group structure. We require a stronger condition, one that ensures that corresponding elements of the two groups are isometries of

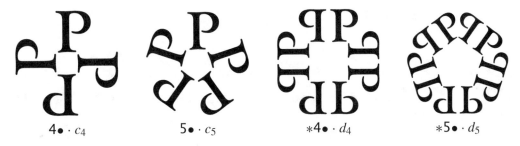

$4\bullet \cdot c_4$ $5\bullet \cdot c_5$ $*4\bullet \cdot d_4$ $*5\bullet \cdot d_5$

Figure 3.1: Examples of figures in the Euclidean plane with cyclic or dihedral symmetry. Each figure is labeled with both its orbifold signature and the name of the abstract group given by its symmetries. (A few notes about the orbifold signatures used in this and subsequent figures are given in Section 3.6.3.)

the same kind. With this requirement in mind, we say that two discrete symmetry groups G and H are *equivalent* if there is an affine transformation A of the plane such that $H = AGA^{-1} = \{A\sigma A^{-1} | \sigma \in G\}$. Intuitively, every symmetry of H can be carried out by temporarily transforming the plane in such a way that the symmetries of G apply, applying a suitable σ in G, and then transforming back.

The set of discrete symmetry groups in the plane is completely understood, and, up to the equivalence defined above, they fall naturally into families. At the topmost level, we classify a discrete group G based on the kinds of translations it contains:

- If G contains no translations at all, then G is either c_n (the *cyclic group of order n*) or d_n (the *dihedral group of order* $2n$). The former is the symmetry group of an n-armed swastika; it consists of n rotations around a single point by integer multiples of $2\pi/n$. The latter is the symmetry group of a regular polygon with n sides; it consists of the rotations in c_n together with n evenly spaced lines of reflection that pass through the centre of rotation. Examples of these symmetry groups are illustrated in Figure 3.1.

- If G contains a family of translations that are all parallel, then G (and any shape with G as its symmetry group) must repeat along an infinite strip because a translation can be iterated any number of times. A shape with this kind of symmetry is called a *frieze pattern*, and its group is a *frieze group*. There are exactly seven inequivalent types of frieze group. Figure 3.2 shows patterns belonging to each of the frieze groups.

- The remaining case is when G contains translations in two linearly independent directions. In this case, a shape with G as its symmetry group must fill the entire plane. The shape is then called a *wallpaper pattern*, and its group a *wallpaper group*. There are 17 types of wallpaper group. They are illustrated in Figure 3.3. These groups contain a large number of symmetries that interact with each other; Figure 3.4 shows a sample wallpaper pattern annotated with symbols that show its symmetries.

The simplest wallpaper patterns are those with signature **o**. These patterns have only translations as symmetries. Note that the symmetry groups of all other wallpaper patterns must contain the symmetries of **o** as a subgroup.

Because of the way wallpaper patterns repeat at regular intervals across the entire plane, they are also referred to as *periodic*. We will favour this term when discussing tilings with wallpaper symmetry.

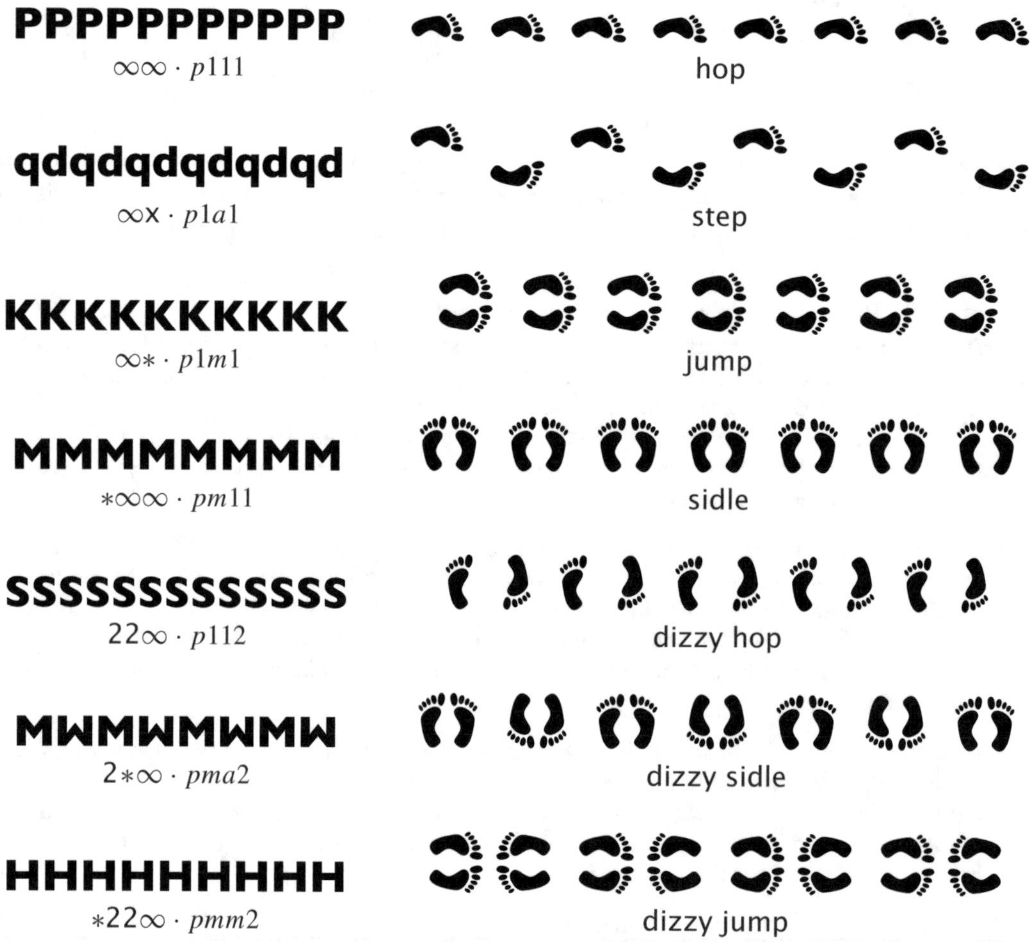

PPPPPPPPPP
$\infty\infty \cdot p111$

hop

qdqdqdqdqdqd
$\infty\mathrm{x} \cdot p1a1$

step

KKKKKKKKK
$\infty* \cdot p1m1$

jump

MMMMMMMM
$*\infty\infty \cdot pm11$

sidle

SSSSSSSSSSSSS
$22\infty \cdot p112$

dizzy hop

MWMWMWMW
$2*\infty \cdot pma2$

dizzy sidle

HHHHHHHHH
$*22\infty \cdot pmm2$

dizzy jump

Figure 3.2: Examples of figures belonging to each of the seven frieze groups. The drawing on the left of each row is formed from letters, and is labeled both with its orbifold signature and the older crystallographic notation. The design on the right is formed from footprints and labeled with names suggested by Conway.

Figure 3.3: Examples of patterns belonging to the 17 wallpaper groups. Each drawing is labelled first with its orbifold signature then with its crystallographic notation.

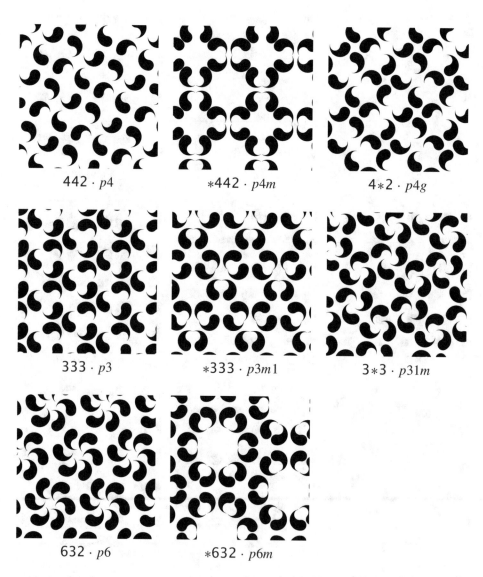

Figure 3.3: Continued.

Figure 3.4: A visualization of the symmetries of a pattern of type 4∗2 (as given in Figure 3.3). The rhombs and squares indicate centres of twofold and fourfold rotation, respectively. The dotted lines are lines of mirror reflection. The dashed lines are lines of glide reflection (the length of the glide reflection's translational component is the distance between adjacent dashed lines). The grey arrows represent two translational symmetries from which all others can be generated.

3.3 FACTORING OUT REPETITION

Symmetry implies redundancy. If a drawing in the plane possesses a mirror symmetry, we need only keep half of the drawing; the other half can be reconstructed from it if we know where the mirror is. More generally, a drawing's symmetries will partition the plane into disjoint orbits. If we know the behaviour of the drawing at just one point in each orbit, we can combine this information with the drawing's symmetry group to reconstruct the original drawing.

We should endeavour to select a representative from each orbit in a way that produces a convenient set as a result. In the spirit of Grünbaum and Shephard [24, Section 1.6], we arrive at the following definition:

Definition 3.1. Fundamental Region. A set U is a *fundamental region* of a discrete symmetry group G if:

1. U is a connected set with non-empty interior;
2. No two points of U belong to the same orbit; and

3. U is as big as possible, in the sense that there does not exist a strict superset of U satisfying the first two properties.

The definition of U as connected with non-empty interior comes across as overly technical. It would be preferable to be able to describe U more simply, for instance, by defining it as either an open or closed topological disc. However, in general, it cannot be either. For example, in the case of wallpaper patterns of type o, the fundamental region has the form of a parallelogram but includes only half of its boundary. In practice, this mathematical detail is an inconvenient distraction. We simply work with the closure of the fundamental region, and recognize that it contains a negligible amount of redundant information. This situation is reminiscent of the intersections that occur in a tiling when tiles are taken to be closed (indeed, the closure of a wallpaper group's fundamental region will always be the prototile of a monohedral tiling of the plane).

In general, a symmetry group may have many possible fundamental regions. However, the possibilities grow increasingly limited for highly symmetric groups because the interior of a fundamental region can never contain an axis of rotational symmetry or intersect a line of reflection.

3.4   PERIODIC REPLICATION

It is the factoring of a pattern into a fundamental region and a symmetry group that makes computers such ideal tools for manipulating patterns. Given a drawing restricted to the fundamental region of a periodic group, the computer can iteratively apply symmetries to that drawing to fill any bounded container region with a subset of an overall wallpaper pattern. (This replication will still work if the drawing wanders outside of the fundamental region, but the results might be harder to predict and control.)

Let us consider more closely the problem of writing a replication algorithm for periodic patterns. We begin with the relatively simple case of patterns of type o, whose symmetries consist purely of translations. Every translation is associated with a vector in the plane. We identify two such vectors $\vec{v}_1$ and $\vec{v}_2$ that are linearly independent and as short as possible among all vectors parallel to them. In this case, the symmetry group consists entirely of translations by vectors $a\vec{v}_1 + b\vec{v}_2$, for all integers a and b. For any point p, a suitable fundamental region for this symmetry group is a parallelogram with vertices $\{p, p + \vec{v}_1, p + \vec{v}_1 + \vec{v}_2, p + \vec{v}_2\}$ (in a computer implementation, it is easiest to choose $p = (0, 0)$). This region is sometimes called a *period parallelogram*. Let us further suppose that we are provided with a subroutine DRAW($\vec{v}$) that draws a copy of the fundamental region translated by any vector $\vec{v}$. If R is a rectangular region of the plane that we wish to fill with a portion of the pattern, we must then decide for which pairs a, b of integers we should invoke DRAW($a\vec{v}_1 + b\vec{v}_2$).

An elegant solution can be found by performing a change of basis into a coordinate frame with origin p and basis vectors $\vec{v}_1$ and $\vec{v}_2$. This change of basis can be interpreted as an affine transformation that projects copies of the fundamental region into a grid of unit squares, and turns R into a parallelogram. In this basis, the regions to draw correspond to those squares that intersect R. In other words, we need only rasterize the representation of R in this frame; the coordinates of the pixels taken to overlap R are precisely the pairs (a, b) needed above. See Figure 3.5 for a visualization of this process.

A similar approach can be taken in which we precompute a raster image of the period parallelogram, and rely on texture mapping hardware to perform the replication. The change of basis above can be used to map any period parallelogram into a square, meaning that the portion of the pattern within that region can conveniently be rasterized into a square texture. Transforming texture coordinates at runtime by that same change of basis will reverse the distortion and reconstruct the original pattern, as illustrated in Figure 3.6.

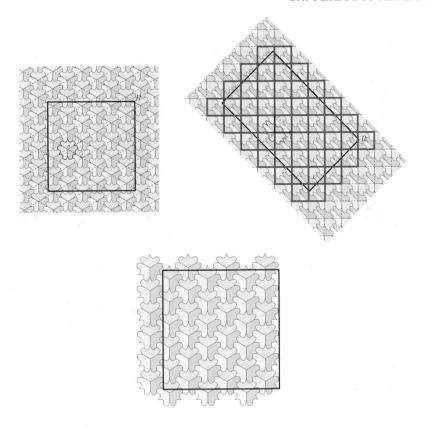

Figure 3.5: The replication algorithm for periodic tilings. The top left image shows the tiling to be replicated, with a superimposed black square representing the desired viewing region R. The dashed lines delineate period parallelograms of the tiling, based on vectors $\vec{v}_1$ and $\vec{v}_2$ and point p. A black outline shows the three tiles that make up a translational unit. In the top right image, the whole diagram is shown in a coordinate system where $\{p, \vec{v}_1, \vec{v}_2\}$ is an orthonormal frame. In this coordinate system, period parallelograms are lattice squares, and a rasterization algorithm can be used to choose the squares that overlap the viewing region. The chosen translational units are drawn in the untransformed image at the bottom. This algorithm can leave part of the viewing region unfilled (as seen in the bottom image) because the tiles chosen to make up a translational unit do not exactly fill the period parallelogram.

The foregoing discussion appears to apply only to **o**, but it can easily be extended to cover the other wallpaper groups. The definition of period parallelogram can be applied to any other wallpaper group G, simply by recognizing that G must contain **o** as a subgroup. We disregard all other symmetries and choose a period parallelogram based on the translations of G. Generally, this parallelogram will contain multiple fundamental regions' worth of information, but this overhead is minimal relative to the simplicity of the resulting algorithm. Alternatively, we can identify a set of rigid motions that compose a

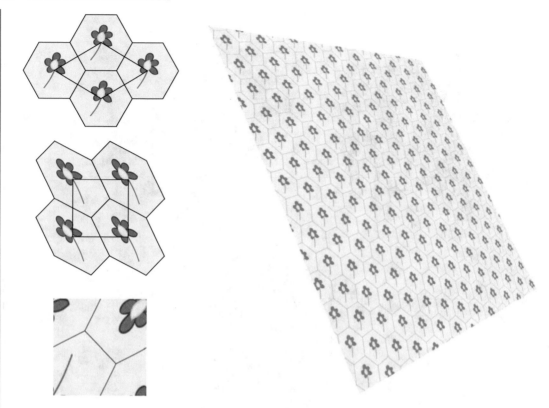

Figure 3.6: An example of periodic replication handled via texture mapping. A portion of a periodic pattern appears on the top left, with a single period parallelogram outlined. The middle left image shows the same drawing transformed so that the period parallelogram is a square. The square image used for texture mapping is shown in the lower left. (It is flipped top to bottom because of the difference between world coordinates and image coordinates.) With suitably chosen texture coordinates the distortion is eliminated on the right, and the pattern is replicated onto 3D geometry.

period parallelogram from copies of a fundamental region. Then, for each translation at which the period parallelogram is to be drawn, we compose that translation with each of the rigid motions and draw the fundamental region.

For an excellent and thorough presentation of computational tools for analyzing and synthesizing wallpaper patterns, see the paper by Ostromoukhov [40].

3.5   SYMMETRIES OF TILINGS

Obviously, we would like these notions of symmetry to apply to tilings as well. However, we need to be careful because a tiling is not a single shape but a countably infinite arrangement of shapes; a symmetry should be thought of as operating on whole tiles rather than points in the plane. We will say that a transformation σ is a *symmetry of a tiling* $\mathcal{T} = \{T_1, T_2, \ldots\}$ if σ is an isometry and for every i, $\sigma(T_i) = T_j$ for some j. That is, a symmetry of a tiling is an isometry that permutes the tiles. It is also possible to take a pictorial approach, and turn the tiling into a drawing S from which the symmetries may be extracted as previously explained in Section 3.2. In that case, the simplest method is to let S be the frontier of the tiling; equivalently, we can use the union of the interiors of the tiles (i.e., the complement of the frontier).

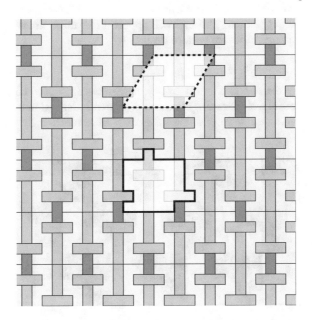

Figure 3.7:   A trihedral periodic tiling. An example of a period parallelogram is outlined with a dashed line. Beneath it, a translational unit made from a patch of tiles is shown. The S-shaped tile occurs in one direct and one reflected aspect, the 1×3 rectangle occurs in two aspects, and the 1×2 rectangle occurs in a single aspect.

In a periodic tiling, a period parallelogram may slice unpredictably through tiles. It is frequently more convenient to express the translational symmetries of a tiling via a unit consisting of a union of whole tiles. In any periodic tiling, we can always find a finite set of tiles whose union forms a fundamental region of the tiling's translational subgroup. We refer to any such set of tiles as a *translational unit* of the tiling. Note that the existence of translational units implies that every periodic tiling is necessarily k-hedral for some finite k. Furthermore, within a translational unit, each of the k prototiles can occur in only finitely many orientations and reflected orientations. We refer to these orientations collectively as the prototile's *aspects*, and distinguish the *direct aspects* from the *reflected aspects* when necessary. Figure 3.7 shows a periodic tiling for which the various prototiles occur in different numbers of aspects.

The patch-based representation of a periodic tiling is useful when tiles must be drawn explicitly via a subroutine rather than sampled from a texture. However, this method is only truly correct when the patch is itself a parallelogram. In practice, it is usually necessary to add one or more rings of additional translational units around the rasterized pixels of R. This adjustment is usually sufficient for rendering, though no matter how many rings are chosen in advance, there will exist a tiling for which that number is insufficient to draw all the tiles that intersect R.

3.6   OTHER FORMS OF SYMMETRY

Since the development of symmetry theory towards the end of the nineteenth century, the basic theory has been expanded in many directions. I close this chapter with a brief discussion of some of the ways symmetry theory has been augmented. The first will appear a few more times in this book; the others lead to other fascinating opportunities for pattern design.

3.6.1   COLOUR SYMMETRY

The shapes from which we developed a theory of symmetries were taken to be subsets of the plane. A shape S can be seen in terms of a *characteristic function* $\chi_S(p)$ that is defined to be 1 when p is in S, and 0 otherwise. Viewed this way, a symmetry of a shape S is then an isometry σ for which $\chi(p) = \chi(\sigma(p))$ for all points p.

Figure 3.8: An example of counterchange symmetry. When considered in isolation, the black pattern has symmetry group **o**. However, we can expand our understanding of symmetry to incorporate the possibility of colour reversal, in which case the pattern can be seen also to have halfturns as symmetries.

In some cases it seems unreasonable to distinguish so sharply between a set S as the "foreground" and everything else as the "background". Consider, for example, the drawing in Figure 3.8. If we restrict our attention to the set of black points in the plane, we find that this set has only translational symmetries, whence the design must belong to wallpaper group **o**. However, this analysis ignores the obvious correspondence between the foreground and background, which can be seen as congruent. Indeed, a 180 degree rotation about any of the points labelled with discs in the figure (and many others besides) will perfectly interchange black and white forms. If we consider *all* symmetries, regardless of whether they interchange colours (or, in this case, if we simply disregard colour and treat the black and white forms as tiles), the design belongs to wallpaper group 2222. We should ask whether there is a mechanism that can take colour into account, one that reconciles the coloured and uncoloured viewpoints.

The solution is to construct a new set of motions that may be used as symmetries, one that takes into account colour reversals. We define every motion as a pair (α, ρ), where α is an isometry of the plane and ρ is a permutation of the set $\{0, 1\}$ (that is, ρ is either the identity function or the function that swaps 0 and 1). Given a shape S with characteristic function χ_S, the effect of this motion is to produce a new shape S' whose characteristic function is defined by $\chi_{S'}(p) = \rho(\chi(\alpha(p)))$. In other words, we apply a rigid motion to S as before, but afterwards we optionally exchange foreground and background. Two motions (α_1, ρ_1) and (α_2, ρ_2) can easily be composed by composing the isometries and permutations separately. It follows that a complete theory of discrete symmetry groups can be developed from this richer set of motions, one that can account for patterns made from alternating black and white forms. These groups, and the shapes that have them as symmetries, are called *counterchange groups* and *counterchange patterns*, respectively.

Counterchange symmetry can naturally be extended to any number k of symbolic colours. We define a *coloured image* to be a function χ from the plane to the set $\{1, \ldots, k\}$. Here, every motion (α, ρ) combines an isometry with a permutation of $\{1, \ldots, k\}$. This motion is a *colour symmetry* of an image χ if $\chi(p) = \rho(\chi(\alpha(p)))$ for all points p. This means that the coloured shape χ can be brought into correspondence with itself by transforming it via a rigid motion and then relabelling all the colours.

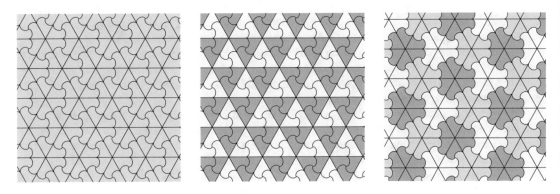

Figure 3.9: Two examples of perfect colourings. The uncoloured tiling (of isohedral type IH30, using the language of Chapter 5) is shown on the left. Perfect colourings using two and three colours are given in the middle and on the right.

A little bit more can be said when considering colour symmetries of tilings instead of general images. A k-*colouring* of a tiling $\mathcal{T} = \{T_1, T_2, \ldots\}$ is a function c over the natural numbers that assigns a colour $c(i)$ in the range $1, \ldots, k$ to each tile T_i (with the assumption that each colour label is used at least once). As before, every colour symmetry of the tiling is a combination of an isometry of the plane with a permutation of the colours. Obviously, every isometry that appears in a colour symmetry is itself a symmetry of the uncoloured tiling, but the converse need not be true. It may not be possible to find a single permutation of the colours that accounts for the change introduced by a given uncoloured symmetry. If every symmetry of the uncoloured tiling can be combined with a permutation to form a colour symmetry, then the colouring $c(i)$ is said to be a *perfect colouring*. Figure 3.9 shows a tiling together with two different perfect colourings with different numbers of colours.

M.C. Escher studied colourings of tilings in depth while preparing his notebook drawings. He paid great attention to the question of colouring, expressing as a clear objective that adjacent tiles should have contrasting colours to better distinguish them from each other [14]. In general, he aimed to achieve this contrast with a minimal number of colours. Yet his intuition seems to have guided him to the perfect colourings, in some cases choosing a perfect colouring with more colours over a non-perfect one with fewer. A clear example is Symmetry Drawing 20 [45, Page 131], where a tiling coloured perfectly by four colours is accompanied by a note mentioning that three would have sufficed to distinguish adjacent tiles. Shephard points out that for this tiling, no perfect colouring is possible with only three colours [47]. Escher intuited that a fourth colour allowed for a more regular colouring.

Escher's understanding of the compatibility between a tiling's symmetries and its colouring pre-dated the development of a formal theory of colour symmetry, and to some extent set that development in motion [44]. For while a small amount of mathematical work had been done on the subject previously, it was when the crystallography community became aware of Escher's tessellations that they understood how much of a theory there was to be had, and they were provided with a rich library of illustrations from which to build that theory.

3.6.2   SYMMETRY IN OTHER SPACES

As mentioned in the introduction, the concept of symmetry extends naturally to many other spaces, both geometric and abstract. It is easy to see that in any metric space (a set endowed with a means of measuring distances), it is possible to use the set of isometries of that space as a basis for understanding the symmetries of shapes. This observation leads immediately to enumerations of discrete symmetry groups in three- and higher-dimensional Euclidean space, as well as in non-Euclidean spaces such as the surface of a sphere and the hyperbolic plane. Mathematicians have also studied symmetry groups of more specialized geometric spaces such as cylinders and thin rods.

Another direction in which symmetry may be extended is by adding additional properties to the space, and coupling isometries with modifications of those properties. We have already encountered an example of this variety of extension with colour symmetry. Another is the study of symmetry groups of the two-sided plane, where a figure is taken to have a "front" and "back" that can be distinguished from one another. The classification that results (which is related, but not identical to counterchange symmetry) is useful in understanding patterns in woven fabrics, for instance.

Symmetry is a powerful, but ultimately limited way to account for the structure of a pattern. The problem is that symmetry can only characterize *global* repetition. There exist many patterns, designed by artists and by mathematicians, which intuitively have a great deal of repetition, but relatively little symmetry. There have been some criticisms of the "cult of symmetry" for its acceptance of symmetry theory as the final word on patterns [22], and attempts have been made to develop more comprehensive theories of repetition.

3.6.3   ORBIFOLDS

Orbifolds represent an alternate way to think about the mathematical basis for symmetries. They do not allow us to create new pattern types that were previously unavailable, nor is it likely that they would lead to a drastic reformulation of the algorithms used to render symmetric drawings in software. However, they provide an attractive, rigorous, intuitive infrastructure from which we can classify the discrete symmetry groups of this chapter and understand why they have the properties that they do.

For many years, information about orbifolds was available only in technical mathematical papers or by word of mouth in a chain of dissemination that flowed outward from John Conway. Happily, this situation is finally remedied with the publication of the new text by Conway et al. [6]. They offer a complete treatment of symmetric patterns in the plane based on orbifolds.

Recall from Section 3.3 that a fundamental region is a set that contains a representative from every one of a symmetry group's orbits. More formally, given a symmetry group we can construct an equivalence relation $\sim$ between points in the plane. We say that $p \sim q$ when p and q are in the same orbit (i.e., when there exists a symmetry that maps p to q). The quotient of the plane by this equivalence relation then acts very much like a fundamental region. But this quotient can be shown to have additional topological and geometric properties that can be described succinctly. For example, for symmetry group **o** consisting of translations, the quotient can be seen as a parallelogram in which opposite edges are identified—in other words, a torus.

For a general discrete symmetry group in the plane, the quotient takes the form of a two-dimensional manifold. The manifold may or may not have a boundary, and may have additional points (called *cone points*) that correspond to orbits that are centres of rotational symmetry. It is this decorated manifold that is referred to as an *orbifold*.

Every discrete symmetry group has a corresponding orbifold. If two groups are equivalent (in the sense given in Section 3.2), then their orbifolds are topologically equivalent. The orbifold is then a concise, natural representation of a symmetry group.

An orbifold can be described by enumerating its features: the presence or absence of a boundary, the number and types of cone points, and so on. This enumeration results in what is called an *orbifold signature*. The signatures function well as a set of names for the symmetry groups, displacing the older crystallographic names. They are attractive because the features of the symmetry group can be read directly from the notation. Indeed, the "magic theorem" presented by Conway et al. uses the orbifold signature to prove that there are exactly 17 wallpaper groups.

EXERCISES

1. Using geometry or linear algebra, prove that the composition of any two reflections is either a rotation or a translation.

2. Let $\triangle ABC$ and $\triangle A'B'C'$ be two congruent triangles in the plane. Prove that there is exactly one isometry that maps $\triangle ABC$ to $\triangle A'B'C'$ (and hence that every isometry is completely determined by its behaviour on one triangle).

3. Prove that every isometry is a product of at most three reflections. (Hint: use the result of the previous question. Pick a triangle in the plane and find a reflection-based way to map it onto its image under the isometry.)

4. Let us investigate the symmetry properties of quadrilaterals.

   (a) Enumerate the symmetries of a square.

   (b) Consider all possible subsets of the symmetries of the square. For some of those subsets (including the set itself), there exist unmarked quadrilaterals possessing those symmetries and no others. Identify those subsets by drawing a representative quadrilateral for each. Arrange your drawings into a directed graph so that there is a path from quadrilateral A to quadrilateral B if A's symmetries are a superset of B's.

Note: we don't want to classify quadrilaterals just by their group structure. There may be visually distinct quadrilaterals with identical symmetry groups.

(c) Prove that there does not exist an unmarked quadrilateral with symmetry group c_4.

5. If you fold a strip of paper into a zig-zag accordion, cut a shape out of it, and unfold the strip, you obtain a frieze pattern of type $*\infty\infty$ (a "sidle" pattern), made from repeated copies of the shape. This folding and cutting method is the standard way to create paper dolls.

Figure out how to create paper dolls belonging to the other six frieze types. The cutting step should still involve cutting out only one copy of a shape; the key is to devise a folding method that "encodes" the desired symmetries.

6. The best way to become comfortable with the frieze and wallpaper groups is to study actual patterns. To some extent you can determine a pattern's symmetry group simply by comparing it to the samples in Figure 3.3. A more principled approach is to use a flowchart with branches based on identifying classes of symmetries in a pattern [51].

(a) Explore the many patterns reproduced by Owen Jones in his classic work *The Grammar of Ornament* [29] (an online copy can be found at
`http://digital.library.wisc.edu/1711.dl/DLDecArts.GramOrnJones`).

Try to find at least one example of each frieze and wallpaper pattern type. Pay attention to what features or imperfections in the drawing must be ignored in order to make the symmetries work.

(b) Find and photograph examples of frieze and wallpaper patterns in your neighbourhood or region. Obviously, the prevalence of interesting examples depends on the architectural styles that dominate in your area; however, even humble bricks occur in a variety of wallpaper patterns [6, Page 42].

7. Although the frieze patterns are visually distinct, they do not give rise to seven different group structures. Find all the pairs of frieze pattern types whose groups are isomorphic.

8. Symmetry theory and tiling theory occasionally produce seemingly arbitrary numbers. Why do there happen to be exactly seven frieze groups or 17 wallpaper groups? This question can be answered by recognizing that certain combinations of symmetries force others into existence, as we shall see here for the frieze groups.

(a) In any frieze pattern, there are four distinct kinds of symmetries that are possible. What are they?

(b) We might, therefore, hypothesize the existence of *16* frieze groups, one for each subset of these four classes of symmetry. Create a table with one column for each kind of symmetry, and 16 rows, one for each boolean assignment to the columns. Identify the seven frieze groups as rows in the table. Then show why the remaining combinations cannot exist.

9. Prove that the interior of a fundamental region can never contain the centre of a rotational symmetry or part of a line of reflection.

10. Write a program that implements the periodic replication algorithm explained at the beginning of Section 3.4. Your program should accept as input a file consisting of two translation vectors and a list of simple geometric primitives (for example, coloured polygons) confined to a period parallelogram defined by the translation vectors. You should display a window containing a rectangle that defines the region R mentioned in the algorithm. The user should be able to translate, scale and rotate the rectangle; the program should respond by replicating the pattern within the transformed R. Make sure that the algorithm is drawing just enough period parallelograms to cover R.

11. Write a program that implements the replication algorithm based on texture mapping. The program should accept a square texture image containing the re-mapped period parallelogram of a pattern, together with the translational symmetry vectors of the original pattern. Use the texture to render the pattern onto a simple 3D primitive such as a triangle or square. The user interface should include the ability to rotate the geometry in 3D, as well as the ability to translate, rotate and scale the pattern relative to the geometry.

*12. For a more complex challenge, modify the texture mapping algorithm of the previous question to work with random-access vector textures, as in the work of Qin et al. [43] and Nehab and Hoppe [39]. It is necessary to modify these algorithms to perform correct antialiasing across boundaries between adjacent copies of the period parallelogram, and to deal with extreme texture minification for distant parts of the pattern.

13. In Section 3.5, the periodic replication algorithm is said to be flawed when a tiling's translational unit is not itself a period parallelogram. It is suggested that the algorithm be modified so that one or more rings of translational units are added around those determined by the algorithm.

    Show that no matter how many rings are added, there will always exist a tiling for which that number is inadequate. You can do so by finding a continuous family of shapes such that:

    i. Each shape is the prototile of a periodic, monohedral tiling of the plane;

    ii. One copy of the prototile is a translational unit for the tiling; and

    iii. For any n there is a shape in the family for which any period parallelogram will overlap at least n tiles.

14. Section 3.6.1 offers a definition for symmetries of a coloured image. However, the notion of a perfect colouring is only introduced in the context of coloured tilings, rather than coloured images in general. Examine the definition of perfect colouring and determine why it cannot be applied to coloured images. What is the simplest class of object in the plane to which the notion of perfect colouring might apply?

CHAPTER 4

Tilings by Polygons

In most cases, we will be interested in tilings by polygons. All of the important features of tilings we will encounter in this book are adequately explained in terms of polygonal tilings. Even when we wish to render tiles with curves edges, those edges will likely be represented as piecewise-linear paths at some stage in the rendering pipeline.

In a polygonal tiling, there can be some confusion between the tiling vertices and edges as described in Section 2.2 and the vertices and edges of individual polygons. To avoid confusion, we refer to the latter features when necessary as *shape vertices* and *shape edges*. Shape vertices and edges are properties of tiles in isolation; tiling vertices and edges are topological properties of the assembled tiling.

When a polygonal tile T is placed in a tiling $\mathcal{T}$, its boundary can be decomposed into an alternating sequence of tiling vertices and tiling edges. This decomposition need not coincide with the tile's description in terms of shape vertices and shape edges. An example where the two sets of vertices and edges are not identical is shown in Figure 4.1. When the two sets of features coincide (that is, when every tiling vertex is a shape vertex and vice versa, the tiling is called *edge-to-edge*. In an edge-to-edge tiling we can simply speak of vertices and edges unambiguously.

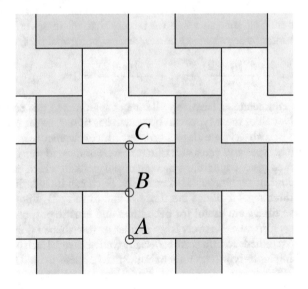

Figure 4.1: The features of a polygonal tiling. For the highlighted tile, A is a shape vertex but not a tiling vertex, B is a tiling vertex but not a shape vertex, and C is both a tiling vertex and a shape vertex. Because the tiling vertices and shape vertices do not coincide, the tiling is not edge-to-edge.

4.1   REGULAR AND UNIFORM TILINGS

A *regular tiling* is an edge-to-edge monohedral tiling of the plane by congruent regular polygons. Recall that the interior angle of a regular n-sided polygon is $\pi(n-2)/n$. If an integral number of such polygons are to meet around every vertex, then we require that $k\pi(n-2)/n = 2\pi$ for some integer k, or that $2n/(n-2)$ is an integer. It is easy to see that this requirement can hold only for $n = 3, 4$ or 6, meaning that the only regular tilings of the Euclidean plane are the familiar ones by squares, equilateral triangles, and regular hexagons.

Let us relax the condition that the tiling be monohedral, and consider the more general case of an edge-to-edge tiling by regular polygons. In such a tiling, every tiling vertex will be surrounded by some collection of regular polygons, all of the same side length. This local arrangement can be encoded by enumerating the numbers of sides in the tiles encountered in a loop around a vertex. We obtain a sequence of the form $p_1.p_2 \ldots p_k$. We know from simple considerations of regular polygons that every p_i is an integer greater than or equal to three, and that $3 \leq k \leq 6$. For example, in the regular tiling by squares, every vertex is surrounded by four square tiles, which we denote by saying that every vertex is of *type* $4.4.4.4$. We will typically abbreviate vertex types using exponentiation; the regular tilings by squares, triangles and hexagons can then be said to have vertex types 4^4, 3^6 and 6^3, respectively. Two vertex types are considered equivalent if they can be made to coincide via a cyclic shift and/or a reversal; or, equivalently, if the corresponding arrangements of regular polygons around two points are congruent.

In full generality, we can say little about the structure of edge-to-edge tilings by regular polygons. Even when limited to tilings consisting of squares and equilateral triangles, some experimentation shows that there is more freedom in how tiles are laid out than we can reasonably expect to control mathematically. However, we obtain an interesting analysis when we require that all tiling vertices have the same type $p_1.p_2 \ldots p_k$. Based on the interior angles of the corresponding polygonal tiles, we know we must have

$$\frac{\pi(p_1 - 2)}{p_1} + \ldots + \frac{\pi(p_k - 2)}{p_k} = 2\pi.$$

It is not too hard to enumerate exhaustively all integer solutions to this equation, generating a list of candidate vertex types. Not all of these types can be realized as tilings, in the sense that any attempt to lay out a tiling with vertices of such a type leads to a geometric inconsistency (e.g., overlapping tiles). If we eliminate all the illegal vertex types and construct tilings corresponding to those that remain, we are left with a set of eleven distinct edge-to-edge tilings by regular polygons in which every vertex has the same type. They are commonly known as the *Archimedean tilings*. Each can be identified by placing its vertex type in parentheses. The three regular tilings are those of the form (p^q). Figure 4.2 shows the eleven Archimedean tilings. These tilings are useful for decorative and artistic purposes, and lead naturally to further exploration of tilings formed exclusively from regular or star-shaped polygons.

It happens that the Archimedean tilings are especially well behaved. Each one is periodic, a property that need not follow from the single-type requirement but which happens to hold for the eleven tilings that meet that requirement. Furthermore, the vertices are equivalent to each other in a very strong way—the congruence that carries the polygons around one vertex to the polygons around another will additionally be a symmetry of the tiling. When this happens, we sometimes say that the tiling's symmetries act *transitively* on the vertices. When viewed in the light of these stronger properties, the Archimedean tilings are often also called *uniform*. In the context of their more general discussion of transitivity properties in tilings, Grünbaum and Shepherd also use the term *isogonal* to refer to a tiling in which the vertices are all related by symmetries.

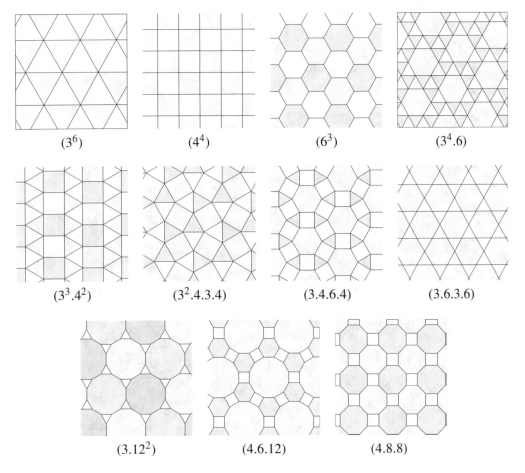

(3^6) (4^4) (6^3) $(3^4.6)$

$(3^3.4^2)$ $(3^2.4.3.4)$ $(3.4.6.4)$ $(3.6.3.6)$

(3.12^2) $(4.6.12)$ $(4.8.8)$

Figure 4.2: The eleven uniform Euclidean tilings, also known as Archimedean tilings. The tiling $(3^4.6)$ occurs in left-handed and right-handed forms.

4.2 LAVES TILINGS

Every Archimedean tiling has a well-defined geometric dual, obtained by placing a tiling vertex in the centre of every regular n-sided tile, and connecting two vertices if their corresponding tiles share an edge. When the original tiling is removed, we are left with a monohedral, edge-to-edge tiling of the plane. Each tile of the dual surrounds a vertex of the original Archimedean tiling, and each vertex of the dual is regular in the sense that the edges adjacent to it are evenly spaced around it. These dual tilings are called the *Laves tilings*, and they are given labels analogous to their Archimedean progenitors. The Laves tilings are depicted in Figure 4.3. They will prove useful in the next section where they serve as a set of "defaults" upon which to describe the more elaborate structure of the isohedral tilings.

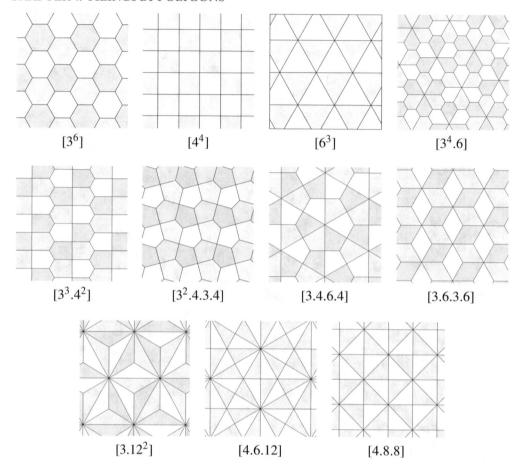

Figure 4.3: The eleven Laves tilings. Each one is the dual of a corresponding Archimedean tiling. The tiling $[3^4.6]$ occurs in left-handed and right-handed forms.

EXERCISES

1. Prove that no monohedral tiling constructed from the prototile of Figure 4.1 can be edge-to-edge.

2. Give an example of an edge-to-edge monohedral tiling whose prototile is non-convex.

3. The fivefold tiling problem asks whether there exists a (normal) tiling of the plane in which every individual tile has fivefold symmetry (that is, every tile has symmetry group c_5). This question remains unsolved [9, 7, 31]. Let us rule out some of the simpler possibilities.

   (a) Prove that regular pentagons cannot tile the plane.

   (b) Prove that regular pentagons and regular pentacles (five-pointed stars with 36° angles at their points) with the same edge length cannot tile the plane together.

(c) Construct the largest patch of tiles you can using just regular pentagons, regular pentacles, and regular decagons, all of the same edge length. Here we must be careful about what is meant by "largest". For a fixed edge length, we measure the size of the patch by the radius of the largest disc it contains.

*(d) Is there anything special about the number five? Can anything be said about the n-fold tiling problem for $n > 6$?

4. Consider a set of prototiles consisting of a regular pentagon together with a rhomb with interior angles of $36°$ and $144°$ and the same edge length as the pentagon. Using these shapes, construct the following tilings:

   (a) A periodic tiling.
   (b) A tiling with d_5 symmetry.
   (c) A tiling with c_5 symmetry.

5. If $p_1.p_2 \ldots p_k$ is a vertex type (an arrangement of k regular polygons of equal edge length in which the ith polygon has p_i sides), why must it be true that $3 \le k \le 6$?

6. Prove that if $p_1.p_2 \ldots p_k$ is a legal vertex type, then $\min\{p_1, \ldots, p_k\} \le 6$.

7. An exercise in *Tilings and Patterns* asks to show that if a tiling by regular polygons contains an octagon, then it must be the Archimedean tiling (4.8^2). As stated, the property to be shown is, in fact, false. Exhibit a tiling by regular polygons that contains an octagon that is not (4.8^2). Then determine the additional assumption necessary to make the statement in the question true.

8. Find all triples of integers p_1, p_2, p_3 that form a legal vertex type. It is only necessary that a single vertex can be surrounded by regular polygons with p_1, p_2 and p_3 sides, not that a complete tiling can be constructed.

9. Show that if the plane can be tiled edge-to-edge by regular polygons in such a way that every vertex is of type $p_1.p_2.p_3$, then p_1, p_2 and p_3 are all even.

10. A *vertex species* is like a vertex type, except that we disregard order. For example, the distinct types $3^2.4.3.4$ and $3^3.4^2$ belong to the same species.

    Write a program to produce a complete enumeration of legal vertex species. This enumeration can serve as a basis for completing a proof that the set of Archimedean tilings in Figure 4.2 is complete.

11. The concept of an Archimedean tiling extends naturally to the sphere and the hyperbolic plane. In particular, a sequence $p_1.p_2 \ldots p_k$ must lie on a sphere if

$$\frac{\pi(p_1 - 2)}{p_1} + \ldots + \frac{\pi(p_k - 2)}{p_k} < 2\pi.$$

Use this inequality to determine the legal Archimedean tilings of the sphere. These can also be realized as polyhedra, in which case they are known as *Archimedean solids*.

12. The Laves tiling $[3^2.4.3.4]$ is sometimes known as the "Cairo tiling" because it is widely used there. Place the prototile from the Cairo tiling so that the edge joining the two 3-valent vertices lies between $(-\frac{1}{2}, 0)$ and $(\frac{1}{2}, 0)$. Give coordinates for the tile's other three vertices.

CHAPTER 5

Isohedral Tilings

The isohedral tilings play a valuable role in art and ornamental design. They correspond to an intuitive notion of "regularity" in monohedral tilings: every tile plays an equivalent role relative to the whole. Despite that constraint, they still permit a wide range of expression. Decorative tilings developed without explicit mathematical knowledge are frequently isohedral. M.C. Escher developed his own "layman's theory" for his regular divisions of the plane [45]; each of his tiling types is equivalent to an isohedral type. On the other hand, the structure of isohedral tilings permits a compact, symbolic description that leads to efficient data structures and algorithms for representing tilings. This combination of expressivity and efficiency allows the isohedral tilings to be used effectively in a variety of computer graphics applications. In this chapter, I present an extended discussion of the mathematical structure of isohedral tilings and computational techniques for manipulating them. The goal is not to derive the enumeration of the isohedral tilings or to prove that the resulting list is complete, but to provide a detailed encoding that suffices to represent and render tilings.

5.1 BASIC DEFINITIONS

Let $\mathcal{T}$ be a tiling. For two congruent tiles T_1 and T_2 in $\mathcal{T}$, there will be at least one rigid motion of the plane that maps T_1 to T_2. A special case occurs when the rigid motion is also a symmetry of the tiling. In this case, when T_1 and T_2 are brought into correspondence, the rest of the tiling will map onto itself as well. We then say that the two tiles are *transitively equivalent*.

Transitive equivalence is an equivalence relation that partitions the tiles into *transitivity classes*. When a tiling has only one transitivity class (and is hence monohedral as well), we call the tiling *isohedral*. More generally, a k-isohedral tiling has k transitivity classes (and may be m-hedral for some $m < k$). An isohedral tiling is one in which a single prototile can cover the entire plane through repeated application of rigid motions from the tiling's symmetry group. In an isohedral tiling, there is effectively no way to tell any tile from any other since the "local views" outward from any two tiles are identical. As a set of simple examples of isohedral tilings, note that all the Laves tilings, shown in Figure 4.3, are isohedral (a fact that will become important in this chapter).

Two tiles in the same transitivity class must obviously be congruent, but the converse need not be true. Figure 5.1 shows a simple monohedral tiling with two transitivity classes. The two classes of tiles can be distinguished by the arrangement of a tile's neighbours around it. This example also demonstrates that being k-isohedral is a property of a tiling, not of its prototiles. The underlying shape used as a prototile in Figure 5.1 also admits an isohedral tiling. We therefore define a k-*anisohedral* set of prototiles as a *set of shapes* that admits a k-isohedral tiling, but no m-isohedral tiling for $m < k$. (We also sometimes refer to a tiling as k-anisohedral when its prototile set is k-anisohedral.) This new definition is truly tied to the prototiles themselves, and places a lower bound on the complexity of the tilings they admit.

In 1900, Hilbert seemed to take it for granted that no anisohedral prototile can exist [24, Section 9.6]. In 1935, however, Heesch demonstrated an anisohedral prototile [25], reproduced in Figure 5.2(a). Since then many more examples have been found of k-anisohedral prototiles. Escher himself drew one 2-anisohedral tessellation, based on a shape introduced to him by Roger Penrose in the form of a wooden puzzle. The search for monohedral, k-anisohedral prototiles remains an active area of research

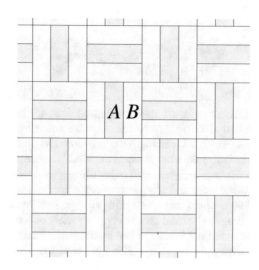

Figure 5.1: The "cheese sandwich tiling", an example of a monohedral tiling that is not isohedral. The tiles fall into two distinct transitivity classes: the "cheese" tiles and "bread" tiles, examples of which are labeled respectively A and B in the tiling. Any two bread tiles correspond via a symmetry, as do any two cheese tiles. But there is no symmetry of the tiling that can make tile A correspond with tile B. This tiling is therefore 2-isohedral. The name of this tiling was suggested by John Berglund.

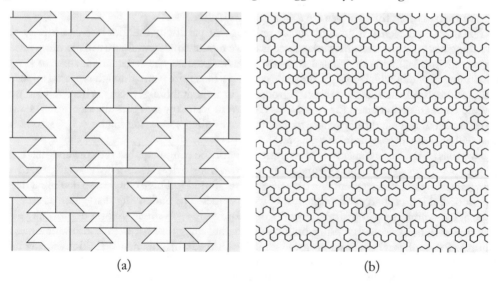

(a)   (b)

Figure 5.2: Anisohedral prototiles. Heesch's 1935 2-anisohedral prototile is shown in (a). The current record holder, a 10-anisohedral 16-hex discovered by Joseph Myers, is shown in (b).

in computational tiling theory [4]. In particular, it is unknown for which values of k such tilings can exist. Joseph Myers initiated a brute-force search for anisohedral polyominoes, polyhexes and polyiamonds, and discovered k-anisohedral prototiles for values of k up to 10 (not including seven!). Figure 5.2(b) shows the 10-anisohedral monstrosity that currently holds the record.

This distinction between k-isohedral and k-anisohedral is a subtle one, and of greater relevance to tiling theory than to computer graphics. However, I wish to emphasize it here because it is important not to confuse properties of tilings with properties of prototiles, particularly in the upcoming section on nonperiodic tilings.

5.2   ISOHEDRAL TILING TYPES

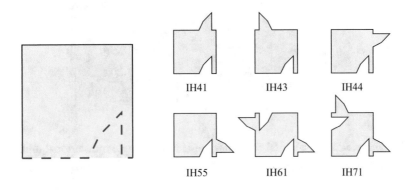

Figure 5.3: An isohedral tiling type imposes a set of adjacency constraints on the tiling edges of a tile. When the bottom edge of the square deforms into the dashed line, the other edges must respond in some way to allow the new shape to tile. The six resulting prototiles tiles here are from six different isohedral types, and show six of the possible responses to the deformation.

By definition, an isohedral tiling is bound by a set of geometric constraints: congruences between tiles must be symmetries of the tiling. Grünbaum and Shephard show that those geometric constraints can be equated with a set of *combinatorial* constraints expressing the adjacency relationships a tile maintains across its edges with its neighbours. They prove that the constraints yield a division of the isohedral tilings into precisely 93 distinct *types* or *families*, referred to individually as IH1, ..., IH93 and collectively as IH [24, Section 6.2]. Each family encodes information about how a tile's shape is constrained by the adjacencies it is forced to maintain with its neighbours. In 12 of these types, the adjacency relationships can be realized only by placing *markings* on tiles that indicate their orientations. We will primarily be concerned with the other 81 types, where the combinatorial structure of the tiling can be expressed geometrically through deformations of the tiling edges. A change to a tiling edge is counterbalanced by deformations in other edges; which edges respond and in what way is dependent on the tiling type, as shown in Figure 5.3. In what follows, we review the classification and notation used with the isohedral tilings.

Recall from Section 2.2 that a tiling's combinatorial structure is an infinite graph that encodes the incidence relationships between the vertices, edges and tiles in a tiling. Furthermore, two tilings with isomorphic combinatorial structures are said to be combinatorially (and topologically) equivalent.

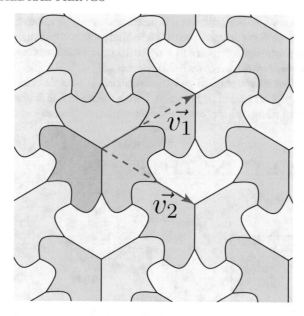

Figure 5.4:   An example of an isohedral tiling of type IH16. A single translational unit of the tiling is shown through the two translation vectors $\vec{v}_1$ and $\vec{v}_2$ and the three coloured aspects.

Grünbaum and Shephard show that combinatorial equivalence partitions the isohedral tilings into eleven classes, referred to as *combinatorial types*, or more commonly as *topological types*. Each topological type has one of the eleven Laves tilings as a distinguished representative, and we name the type using the vertex symbol of the corresponding Laves tiling. For example, Figure 5.4 shows an isohedral tiling of type IH16. We can see that every tile has six tiling vertices, all of valence three, meaning that IH16 is of topological type 3^6.

   Every isohedral tiling is both monohedral and periodic, meaning that its behaviour over the entire plane can be summarized by specifying the aspects of the prototile that make up a translational unit, together with two linearly-independent translation vectors that replicate that unit over the plane. IH16 has three aspects, shown shaded in Figure 5.4. These three tiles comprise one possible translational unit with translation vectors $\vec{v}_1$ and $\vec{v}_2$ as labelled in the figure.

   The adjacency constraints between the tiling edges of a tile are summarized by an *incidence symbol*. Given a rendering of an isohedral tiling, the incidence symbol can be derived in a straightforward manner.

   Figure 5.5 shows five steps in the derivation of an incidence symbol for our sample tiling. To obtain the first part of the incidence symbol, we pick an arbitrary tiling edge as a starting point, assign that edge a single-letter name, and draw an arrow pointing counterclockwise around the tile (Step 1). Then, we copy the edge's label to all other edges of the tile related to it through a symmetry of the tiling (Step 2). Should the edge get mapped to itself with a reversal of direction, it becomes undirected and is given a double-headed arrow. We then proceed counterclockwise around the tile to the next unlabeled edge (if there is one) and repeat the process (Step 3). The first half of the symbol is obtained by reading off the assigned edge names (Step 4). A directed edge is superscripted with a sign indicating the agreement of

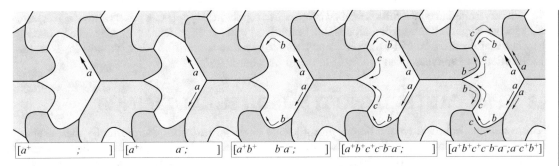

Figure 5.5: Five steps in the derivation of an incidence symbol for a tiling of type IH16.

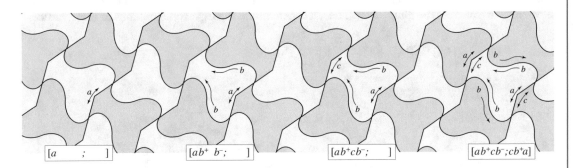

Figure 5.6: Another example of the derivation of an incidence symbol, this time for a tiling of type IH66. In this case, the edges labelled a and c have double-headed arrows, indicating that a symmetry of the tiling (a mirror through the middle of the tile) maps these edges to their own reflections. This symmetry is reflected in the lack of a sign for those edge labels in the incidence symbol.

its arrow with the traversal direction. Here, a plus sign is used for a counterclockwise arrow and a minus sign for a clockwise arrow.

The second half of an incidence symbol records how, for each different label, a tiling edge with that label is related to the corresponding edge of the tile adjacent to it. To derive this part of the symbol, we copy the labeling of the tile to its neighbours (Step 5). Then, for each unique edge letter assigned in the first step, we write down the edge letter adjacent to it in the tiling. If the original edge was directed, we also write down a plus or minus sign, depending on whether edge direction is respectively preserved or reversed across the edge. That is, a plus sign is used if the arrows on the two sides of a tiling edge are pointing in opposite directions, and a minus sign is used otherwise. For the ongoing example, the incidence symbol turns out to be $[a^+b^+c^+c^-b^-a^-; a^-c^+b^+]$. In order to further demonstrate this process, the derivation of a second incidence symbol for a different isohedral type is given in Figure 5.6.

Note that the incidence symbol is not unique; edges can be renamed and a different starting point can be chosen. But it can easily be checked whether two incidence symbols refer to the same isohedral type. Usually we label an isohedral tiling with a canonical incidence symbol by choosing the one that is lexicographically first among all possible incidence symbols for that tiling.

Every isohedral type is fully described in terms of a topological type (i.e., a list of vertex degrees) and an incidence symbol. Enumerating all possible topological types and incidence symbols and eliminating the ones that do not result in valid tilings or that are trivial renamings of other symbols leads to the classification into 93 types given by Grünbaum and Shephard.

5.3   PARAMETERIZING THE ISOHEDRAL TILINGS

Within a single isohedral type, different prototiles are distinguished from each other by their shapes, determined by the positions of the tiling vertices and the shapes of the curves that join them. In order to move from the combinatorial description of isohedral tilings to a geometric one, we must understand how incidence symbols dictate the range of possible prototile shapes for a given isohedral type. We parameterize the space of isohedral tilings by giving, for each type, an *edge shape parameterization* and a *tiling vertex parameterization*. The former encodes the minimal non-redundant geometric information sufficient to reconstruct the tiling edges. The latter determines the legal configurations of tiling vertices.

5.3.1   EDGE SHAPE PARAMETERIZATION

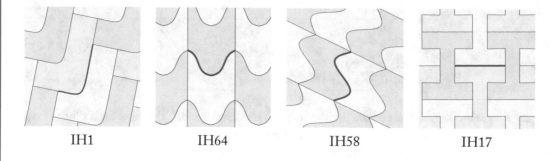

| IH1 | IH64 | IH58 | IH17 |

Figure 5.7: Examples (from left to right) of J, U, S and I edges. In each case, the tiling edge with the given shape is highlighted.

The constraints on the shapes of tiling edges in an isohedral tiling are simple to describe. Although the underlying choice of how to represent a curve is left open, the tiling's symmetries imply a great reduction in the tiling edges' degrees of freedom. These constraints can be extracted directly from the tiling's incidence symbol. We enumerate the four cases for the structure of a tiling edge. For each case, Figure 5.7 shows a tiling with such an edge.

If some directed edge is adjacent to itself without a flip, then a tile's neighbour across that edge is adjacent through a half-turn. This rotation forces the edge shape to itself be symmetric through a half-turn about its centre. We call such an edge an S edge as a visual mnemonic. Only half of an S edge is free; the other half must complete the rotational symmetry. In an incidence symbol, we can identify an S edge as an edge name x that is directed and adjacent to x^{+}.

An undirected edge must look the same starting from either end, meaning it must have a line of mirror symmetry through its midpoint. If an edge name x appears in an incidence symbol without a sign, and is adjacent to some other edge name y that is distinct from x, then x is free to take on any curve with this bilateral symmetry. We call it a U edge. Again, only half of a U edge is free.

If an undirected edge is adjacent to itself, or if a directed edge is adjacent to itself with a change in sign, that edge must have both S symmetry and U symmetry. The only shape that has both is a straight line, leading us to call such an edge an I edge.

The remaining case is a directed edge x that is adjacent to some other directed edge y. Such an edge is free to take on any shape, and we call it a J edge.

Note also that if an edge x is adjacent to an edge y, then the curves drawn for x and y must be congruent (even though they have different names). In this case, an implementation need only represent one of these curves, since the other is entirely constrained to it. Thus, referring back to the derivation presented in Figure 5.5, the tiling edges of IH16 can be summarized by one curve: the shape of the edge labeled b. Edges labeled a are I edges and have no degrees of freedom, and edges labeled c are constrained to b.

Examination of the enumeration of isohedral tiling types reveals that for 21 of the types, all tiling edges must be I edges. For example, the type IH76 corresponds to the regular tiling by squares. The reflection symmetries of that tiling force all edges to be straight. However, 12 of these straight-edged types have a curious property: although all edges must be straight, the symmetries of tiles, as dictated by the incidence symbol, are only a subset of the symmetries of the tiling. Because we cannot distort the edges, the only way to display a tiling with this property is to include a "marking" on each tile, a figure that breaks some of the symmetries that would otherwise be present. We therefore say that these 12 types cannot be realized with unmarked tiles. They are less important in graphical applications—their fixed tile shapes can just as easily be represented via the corresponding (unmarked) Laves tilings.

5.3.2   TILING VERTEX PARAMETERIZATION

Like the shape vertices, tiling vertices cannot move entirely independently of each other. Moving one tiling vertex forces the others to move to preserve tileability. The exact nature of this movement depends on the tiling type in question. The incidence symbol for a tiling type implies a set of constraints on the tiling polygon's edge lengths and interior angles. Any tile of that type will have a tiling polygon that obeys those constraints.

In a constructive model of isohedral tilings, it is not sufficient merely to recognize the constraints on the shape vertices; we need a way to explicitly navigate the space of legal tiling polygons. This section provides explicit parameterizations of the tiling vertex configurations for IH. They can be derived by determining angle and length constraints from the incidence symbols and parameterizing the unconstrained degrees of freedom. In some cases, parameterizations are shared between tiling types: nine tiling types have squares as tiling polygons (implying a parameterization with zero parameters), and seven have parallelograms (implying two parameters). In all, the 93 isohedral types require 43 different parameterizations.

Often, determining the tiling vertex parameterization for an isohedral type is straightforward. It is easy to see, for example, that every isohedral tiling of type IH41 will have symmetry type o, consisting only of translations, and that a single tile can function as a translational unit. In that case, the tiling polygon must be a fundamental region of the tiling's symmetry group, and hence a parallelogram.

To give the flavour of a more complicated parameterization, I offer a sketch of the derivation for IH16 (see Figure 5.9). We begin by placing at least enough tiles to completely surround one central tile, and marking the tiles with the labels from the tiling's incidence symbol. Now consider the situation at the tiling vertex labelled A in the figure. This vertex is surrounded by three copies of the same angle from three different tiles, namely $\angle FAB$, the angle between the a edges. It follows that the tiling polygon must have a 120° angle at that vertex. The same observation applies to vertices C and E. Thus, $\triangle FAB$, $\triangle BCD$, and $\triangle DEF$ are all 120° isosceles triangles. Because these isosceles triangles can be constructed

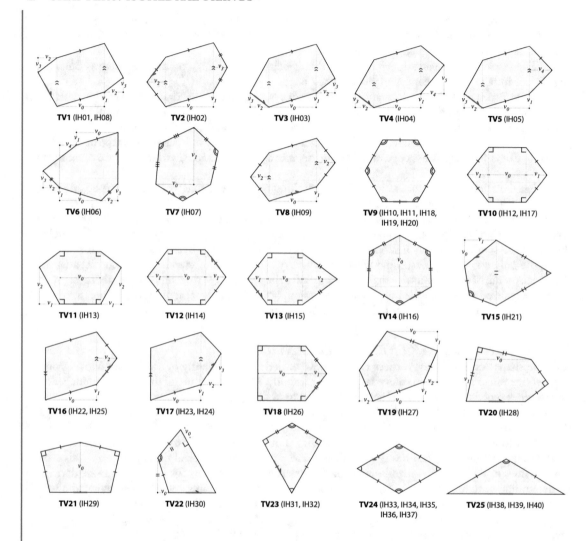

Figure 5.8: The complete set of tiling vertex parameterizations for the isohedral tilings. In each tile, the marked edge is the first edge in the tiling type's incidence symbol (see Appendix A). When that first edge is directed, the marking has an arrowhead. Labelled dotted lines represent parameter values, and are horizontal or vertical (with the exception of one guide line in the diagram for IH30). Since the diagrams are scale independent, distances that do not depend on parameters can be taken to have unit length. Tile edges cut with the same number of short lines have the same length, and edges cut with chevrons are additionally parallel. A single arc, a small square, and a double arc at vertices represent 60°, 90°, and 120° angles, respectively.

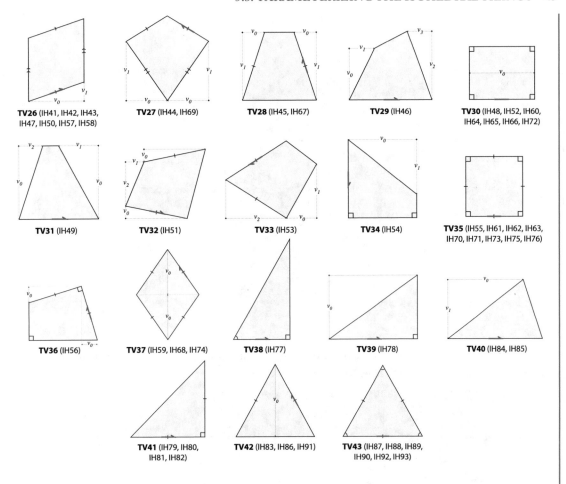

Figure 5.8: The complete set of tiling vertex parameterizations for the isohedral tilings (continued).

given only the edge opposite the 120° angle, the tiling polygon depends entirely on the "skeleton" triangle $\triangle BDF$. Furthermore, the incidence symbol reveals a line of bilateral symmetry in the tile across $\overline{AD}$, forcing $\triangle BDF$ to be isosceles. The only degrees of freedom left in the tiling polygon are the lengths of $\overline{AD}$ and $\overline{BF}$. However, we are not interested in capturing the absolute size of the tiling polygon, merely its shape up to similarity. We can factor out the dependence on scale by fixing $BF = 1$ and keeping just a single parameter: $v_0 \equiv AD$. Figure 5.10 shows tilings of type IH16 that can result from different values of this single parameter.

Diagrams of the tiling vertex parameterization for the isohedral tilings are given in Figures 5.8. These parameterizations can be seen as an elaboration of those provided by Heesch and Kienzle for the 28 Heesch tiling types [26]. Each Heesch type is identical to one of the isohedral types, and for those types the parameterizations coincide. The remaining isohedral types have parameterizations where degrees of freedom are coalesced to yield more symmetric tiling polygons.

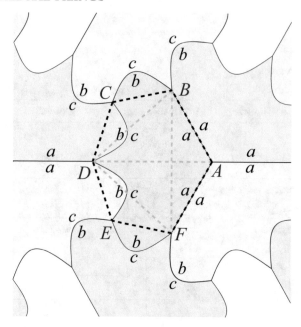

Figure 5.9: The diagram used to establish a tiling vertex parameterization for IH16. For simplicity, the arrows indicating edge direction have been left out of the diagram.

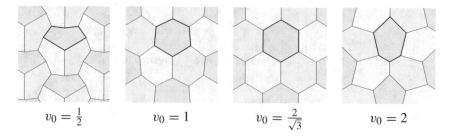

$$v_0 = \frac{1}{2} \qquad v_0 = 1 \qquad v_0 = \frac{2}{\sqrt{3}} \qquad v_0 = 2$$

Figure 5.10: Some examples of IH16 with different values for the single parameter in its tiling vertex parameterization.

5.4   DATA STRUCTURES AND ALGORITHMS FOR IH

In this section, I provide more details on how the edge shape and tiling vertex parameterizations can be developed into a complete implementation for storing, manipulating, and rendering isohedral tilings. These data structures were first devised for the implementation of Tactile, a library I developed to support the Escherization algorithm [33]. The Tactile library has proven useful in many other applications since then.

5.4.1   REPRESENTING TILING VERTEX PARAMETERIZATIONS

It is not hard to believe that the tiling vertex parameterizations in Figure 5.8 can be turned into functions that accept parameters as input and produce the coordinates of tiling vertices as output. The first implementation of Tactile was written this way. However, closer inspection of these parameterizations reveals an important fact: the coordinates of the tiling vertices are all linear in the parameters. That is, if a given isohedral type has a parameterization with parameters $v_1, \ldots, v_n$, then the x and y coordinates of every vertex are the values of expressions of the form $\alpha_1 v_1 + \ldots + \alpha_n v_n + \alpha_{n+1}$ for fixed real coefficients $\alpha_1, \ldots, \alpha_{n+1}$ that depend only on the tiling type.

This observation allows us to give an efficient table-driven implementation of the tiling vertex parameterizations. A parameterization that computes the coordinates of k tiling vertices from n parameters can be expressed as a matrix of coefficients with $2k$ rows and $n + 1$ columns. Let $\vec{r}_i$ denote the ith row of this matrix and let $\vec{v}$ be the vector $(v_1, \ldots, v_n, 1)$. Then the jth tiling vertex will have coordinates $(\vec{r}_{2j-1} \cdot \vec{v}, \vec{r}_{2j} \cdot \vec{v})$ (where $\cdot$ denotes the dot product). We therefore need only derive such a matrix for each of the distinct parameterizations in Figure 5.8 and associate each tiling type with the correct matrix. For example, it can be shown that the parameterization for IH16 as given in Figure 5.8 can be represented by the matrix

$$\begin{bmatrix} 0 & \frac{1}{2} \\ 0 & -\frac{1}{2\sqrt{3}} \\ 0 & 1 \\ 0 & 0 \\ \frac{1}{2\sqrt{3}} & \frac{3}{4} \\ \frac{1}{2} & \frac{1}{4\sqrt{3}} \\ 0 & \frac{1}{2} \\ 1 & 0 \\ -\frac{1}{2\sqrt{3}} & \frac{1}{4} \\ \frac{1}{2} & \frac{1}{4\sqrt{3}} \\ 0 & 0 \\ 0 & 0 \end{bmatrix}$$

5.4.2   COMPUTING TRANSFORMATION MATRICES

Following the discussion in Section 3.5, we will render isohedral tilings by assembling a set of tiles into a translational unit and using the periodic replication algorithm. In order to use this approach, we must compute transformation matrices that position each of the aspects in the translational unit, and suitable vectors that generate all of the tiling's translational symmetries. These matrices and vectors are not purely a property of the isohedral type because they depend on the positions of a prototile's tiling vertices.

I refer to the transformation matrices that move the prototile to the aspects in a translational unit as *aspect transforms*. Suppose that a translational unit for a given tiling consists of tiles $T_1, \ldots, T_m$, one for each of m aspects. Without loss of generality, we can choose the identity as the aspect transform for aspect 1. Our goal is to determine a set of rigid motions to apply to this first aspect to map it to aspects $2, \ldots, m$ in a single translational unit.

Every tile in an isohedral tiling is surrounded in a consistent way by its neighbours, and so for every tiling edge there is a well-defined rigid motion that carries the tile on one side of that edge to the tile on the other side. The motion will either be a half-turn around the edge's center (in the case of an S edge), a reflection across the edge (in the case of an I edge), a glide reflection (in the case of some J edges) or a translation. The kind of transform that applies can be determined from the tiling type's incidence symbol, and the numeric values in the transform matrix depend on the positions of the tiling vertices that delimit the edge. I call such a rigid motion a *hop* across a tile edge. In a tile with n edges, we can label the hops unambiguously as $H_1, \ldots, H_n$.

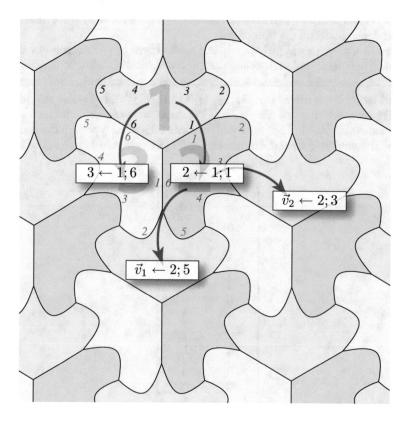

Figure 5.11: A visualization of how aspect transforms and translation vectors are computed for IH16, using rules as described in Section 5.4.2. The edges are numbered as they are given in the incidence symbol. Each arrow represents a single hop, a rigid motion that brings a tile into coincidence with one of its neighbours. The end of every sequence of hops is labeled with a rule for an aspect transform or translation.

Using this information, every aspect transform can be computed by starting from an aspect with a known transform and walking to the desired aspect, accumulating hops across edges along the way. This process can be described by giving a set of symbolic rules, one for each aspect. A rule consists of the name of an aspect, together with a sequence of indices in the range $1, \ldots, n$. The aspect transform can be computed by composing the transform associated with the named aspect with the sequence of hops given by the indices. For example, a rule of the form $4 \leftarrow 2; 4, 3$ implies that the aspect transform for T_4

is the product of the transform for T_2 with the hops H_4 and H_3. For every isohedral type, we can write down an ordered list of rules that will yield a complete set of aspect transforms.

The two translation vectors can be derived using similar rules, but with a twist. As before, we can specify a rule to obtain a transformation matrix. However, that matrix does not necessarily represent a translation, and so we cannot just take the vector to be its translational component. The problem is that the matrix may contain internal symmetries of the tile shape, which were accumulated when composing hops together. Fortunately, we can still extract the translation in a simple way as the vector joining the centroids of the transformed and untransformed tiling vertices. This calculation works because the centroid is independent of internal tile symmetries, operations that merely permute the vertices.

A visualization of how these rules may be applied for the IH16 example is given in Figure 5.11.

5.4.3   COLOURINGS

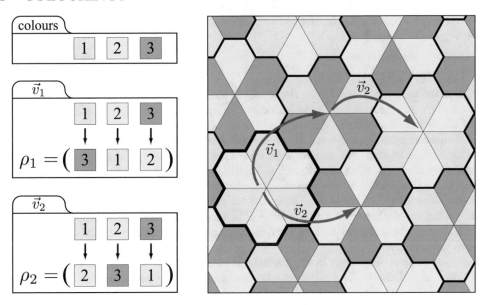

Figure 5.12: A demonstration of how colouring information for an isohedral tiling (for IH21 in this case) is used to apply colours to tiles. The translational units (each containing six aspects) are outlined in bold. There are three symbolic colours, $\{1, 2, 3\}$, and they are associated respectively with gray, pink, and blue. On the left, the permutations for the two translation vectors are indicated by showing with arrows the mapping from original to permuted colours; the permutation's textual description can be read off of the bottom row of this mapping. On the right, the permutations are applied when moving between translational units. The colouring for this tiling can be read from the diagram as $(1\,2\,1\,2\,1\,2)\,(3\,1\,2)\,(2\,3\,1)$.

Just as the isohedral tilings formalize an intuitive notion of regularity, perfect colourings are a natural way to colour tiles in an orderly way. As mentioned in Section 3.6.1, perfect colourings also correspond well to Escher's use of colour in his tessellations. I wish to encode a rule that assigns colours to tiles during rendering.

In a perfectly coloured tiling, every symmetry of the tiling is associated with a permutation of the colours. The actions of all these symmetries can be summarized by giving the permutations associated with the two translation vectors of the tiling and an assignment of colours to the aspects in a single translational unit. Successive translations will permute this default assignment appropriately. A colouring with k colours can then be specified with three sequences of integers of the form $(c_1 \ldots c_m)(p_1 \ldots p_k)(q_1 \ldots q_k)$. The numbers $c_1, \ldots, c_m$ give the assignment of colours to the m aspects in the default translational unit. The p_i and q_i describe permutations of the k colour labels associated with the tiling's translation vectors (the permutation ρ associated with the p_i is simply the function $\rho(j) = p_j$, and likewise for the q_i).

In particular, consider a tiling with translation vectors $\vec{v}_1$ and $\vec{v}_2$ and their associated colour permutations ρ_1 and ρ_2. Then aspect i in the translational unit located at $a\vec{v}_1 + b\vec{v}_2$ will have the colour $\rho_2^b(\rho_1^a(c_i))$. This encoding can in fact express a superset of the perfect colourings, but it is easy to check empirically whether a given colouring is perfect.

In a practical implementation, a colouring will store the information above together with RGB triples for the k colours. An example of a perfectly coloured tiling, together with the permutations that encode it, is given in Figure 5.12.

5.4.4   TILING EDGE SHAPES

In addition to the parameters controlling the locations of the tiling vertices, the aspect transforms, the translation vectors, and the colouring, an isohedral prototile must store descriptions of the non-redundant portion of the tile's outline (which I call the *fundamental edge shapes*).

Each fundamental edge shape is a description of a path starting at $(0, 0)$ and ending at $(1, 0)$. Any curve description can be used here. A natural choice is a piecewise-linear path represented by its vertices. More elaborate paths are possible as well, such as subdivision curves that interpolate their endpoints.

The shape information in the prototile contains a hierarchical model of rigid motions whose leaves are the fundamental edge shapes. The model makes multiple references to fundamental edges to express the redundancy inherent in the tile's outline. To rebuild the tile shape, we apply the tiling vertex parameterization to obtain the positions of the tiling vertices and use the hierarchical model to construct edge shapes between them.

There are at most three levels in the hierarchical model between a fundamental edge shape and a point on the outline of the tile. The first level takes into account the symmetries of U and S edges. Half of the U or S edge comes directly from the fundamental edge. The other half is derived from the first half as needed through rotation or reflection. J edges are passed unmodified through this level, and since I edges are immutable, all tiles can share a single system-wide I edge.

At the next level up, we recognize that edges with different names in the incidence symbol may still have related shapes. In IH16, for example, the edge named b^+ is adjacent to c^+, forcing the two edge shapes to be congruent. In this case, the two edges share the same shape passed up from the level below.

Finally, the topmost level maps the unit interval to an edge of the tiling polygon; this mapping will move an edge shape from its normalized coordinate system into a portion of the tile's outline. At this level, all edges with the same label in the incidence symbol share a lower-level shape object.

5.4.5   ISOHEDRAL TEMPLATES AND PROTOTILES

Putting together all the information associated with an isohedral tiling type, we arrive at a collection of information I call a *template*. An example of the template for IH16 is given in Figure 5.13. The topological type and incidence symbol (Line 1) come directly from Grünbaum and Shephard. (Note that the tiling type is not fully described without the topological type; there is no a priori reason why

1	**IH16**	$3^6 \cdot [a^+b^+c^+c^-b^-a^- ; a^-c^+b^+]$
2		Parameterization TV14
3		Colouring (1 2 3) (1 2 3) (1 2 3)
4		Aspects 1, 2, 3
5		Rules
6		$2 \leftarrow 1; 1$
7		$3 \leftarrow 1; 6$
8		$\vec{v}_1 \leftarrow 2; 5$
9		$\vec{v}_2 \leftarrow 2; 3$

Figure 5.13: An example of an isohedral template, giving all the information associated with the isohedral tiling type IH16.

tilings with different topologies could not have identical incidence symbols, and indeed this situation does arise.) The parameterization in Line 2 is a reference to one of the matrices described in Section 5.4.1. The colouring in Line 3 is a suitable default that uses three or fewer colours. Individual prototiles can override this default with their own colourings. Line 4 gives names to the tiling's aspects, and the names are then used to define rules for the aspect transforms and translation vectors, as described in Section 5.4.2. A complete listing of templates for the 81 unmarked isohedral tiling types, together with a sample tiling belonging to each type, is given in Appendix A. This information has also been available online since 2000, in a file called `isohedral.ih`. It can currently be downloaded from `http://www.cgl.uwaterloo.ca/~csk/projects/escherization/`.

An individual prototile refers to a template, and additionally stores a colouring, a vector of tiling vertex parameters, and a set of fundamental edge shapes. A callback mechanism can help to ensure that changes to upstream information such as tiling vertex parameters can propagate dynamically to the rest of the tiling information that depends on it.

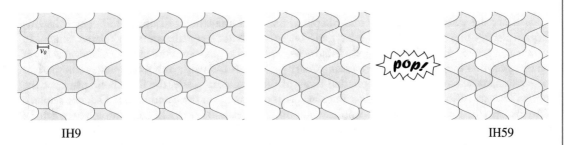

IH9   IH59

Figure 5.14: An example of how a degenerate tiling edge leads to a related tiling of a different isohedral type. As parameter v_0 goes to zero in this IH9 tiling, the tiles deform continuously. But at the instant that v_0 becomes zero, pairs of tiling vertices fuse and the tiling passes through a topological discontinuity to type IH59.

This representation of isohedral tilings suffers from a flaw related to degenerate edges in the tiling polygon. If two consecutive tiling vertices are made to coincide, then the hop across their shared edge is undefined, and any hop-based rules that use the degenerate edge give invalid transforms. In a purely mathematical treatment of the subject, there is no problem because there is no such thing as a degenerate edge in the tiling polygon. As two adjacent tiling vertices merge, they fuse into a single vertex and the tiling as a whole slips into a different (but related) isohedral type. An example is shown in Figure 5.14. The representation given here can manipulate non-degenerate tiles without any difficulty, but it cannot handle these discontinuous transitions. It is not clear that there is sufficient value to justify the complexity of computing these transitions, especially given that they may only be momentary as parameters vary.

In a similar vein, this representation of isohedral prototiles cannot recognize when a prototile's shape happens to acquire more symmetry than is required by the prototile's isohedral type. In this situation, a mathematical classification might place the tile in a different isohedral type that recognizes the extra symmetry. In practice there is no harm in permitting these accidental symmetries; they merely highlight the difference between using incidence symbols for classification versus synthesis. In fact, this ability can be useful. Recall that every isohedral type belongs to one of the eleven topological types exemplified by the Laves tilings of Section 4.2. The Laves tiling can always be chosen as a default prototile shape for all isohedral types with the same topology because it is guaranteed to be within the representational range of those isohedral types. This property was useful in the Escherization algorithm [33] where an optimization algorithm could search through the tile shapes belonging to an isohedral type by starting with the Laves tiling.

5.5   BEYOND ISOHEDRAL TILINGS

Before the isohedral tiling types were enumerated, Heesch and Kienzle developed a family of what are now known as *Heesch tilings* [26, 45]. Each Heesch type is represented by a symbol that encodes the type of adjacency across the tiling edges and the order of rotational symmetry (if any) at the vertices. There are 28 Heesch types, corresponding to the "primitive" isohedral types—those in which no non-trivial symmetry of the tiling maps a tile to itself. They also did not consider prototiles with I edges. In many typical applications the Heesch types are sufficient since the internal symmetries of tiles can always be contrived by choosing edge shapes appropriately. In any case, every Heesch type has a corresponding isohedral type, and so the Heesch tilings can be considered to be subsumed by the discussion in this chapter.

Since the work of Grünbaum and Shephard on the classification of isohedral tilings of the Euclidean plane, other tiling theorists have gone on to search for generalizations to related tilings. In particular, a group led by Dress, Delgado Friedrichs, and Huson pioneered the use of *Delaney symbols* in the study of what they call *combinatorial tiling theory* [10, 28]. A Delaney symbol completely summarizes the combinatorial structure of a k-isohedral tiling of the Euclidean plane, the hyperbolic plane, or the sphere. They can also be generalized to tilings in spaces of dimension three and higher. Delaney symbols form the basis for an efficient software implementation, and Delgado Friedrichs and Huson have created a series of software tools for exploring, rendering, and editing tilings from their combinatorial descriptions (see the Gavrog project at http://gavrog.sourceforge.net/ for more information). With some additional work, Delaney symbols might be used in place of incidence symbols above, offering additional flexibility not described here.

An isohedral tiling's incidence symbol can be seen as a recipe for "sewing up" a prototile along related edges. The process amounts to a generalization of constructing a manifold corresponding to the orbits of a discrete symmetry group—in other words, the sewn-up prototile is an orbifold. Conway et al.

show that this process can be reversed [6, Chapter 15]. Isohedral tilings in the plane, the sphere and the hyperbolic plane can be derived by starting with an orbifold and "cutting it open" into a flat prototile. The combination of an incidence symbol and a topological type then corresponds to an orbifold with a graph drawn on it; the graph's edges indicate where the orbifold should be cut open. Indeed, most of the techniques in this chapter can be recast elegantly in the language of orbifolds. This translation might make some of the tiling type's features more evident. For example, an isohedral type's symmetry group is not encoded anywhere in its incidence symbol or topological type, though it can be derived with some work. That information is more plainly visible in the corresponding orbifold.

EXERCISES

1. Show that there exists a k-isohedral tiling for every positive integer k.

2. Erect a single equilateral triangle on one edge of the 16-hex from Figure 5.2(b) in order to produce a prototile of an isohedral tiling of the plane. This might most easily be done by cutting out copies of the prototile and experimenting with symmetric arrangements of them.

3. One way to prove that the tiling of Figure 5.2(b) is 10-isohedral is to exhibit a patch of ten copies of the prototile that tiles the plane isohedrally. Find such a patch. Then find a translational unit of the tiling.

4. What is the smallest n for which there exists an anisohedral n-omino?

5. (a) Find a shape that is the prototile of four different isohedral tilings, all of different types.

   *(b) What is the largest n for which there exists a shape that is the prototile of isohedral tilings of n different types?

6. Identify two different isohedral types that differ in topology but have identical incidence symbols.

7. Recall from Section 5.3.1 that tiling edges in an isohedral tiling can be classified based on their symmetries as J, U, S and I edges.

   (a) For each of the four edge types, find an isohedral type consisting exclusively of edges of that type.

   (b) For each of the six possible pairs of edge types, find an isohedral type consisting exclusively of edges of those two types.

   (c) Find an isohedral type with edges belonging to three different edge types.

   (d) Show that there cannot exist an isohedral tiling with edges belonging to all four edge types.

8. Derive the tiling vertex parameterizations for IH27, IH28 and IH29.

9. Identify the isohedral types of the tilings shown in Figure 5.15. Disregard colourings and markings.

10. The default colourings given for the isohedral types in Appendix A are all perfect colourings except for a single isohedral type: IH28.

   (a) Show that the colouring provided is not perfect.

   (b) Give a perfect 4-colouring for this tiling in which no two adjacent tiles have the same colour. Use the permutation notation presented in Section 5.4.3.

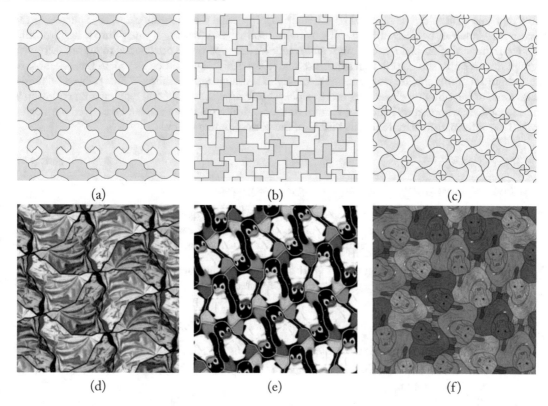

(a)   (b)   (c)

(d)   (e)   (f)

Figure 5.15: Sample isohedral tilings for identification.

    (c) Prove that this tiling cannot be perfectly 3-coloured.

11. The colouring algorithm for isohedral tilings, presented in Section 5.4.3, still suffers from an un-fortunate inefficiency. In the expression $\rho_2^b(\rho_1^a(c_i))$, we must iteratively apply the two permutation functions an unknown number of times, with no upper bound on the values of a and b.

    (a) By precomputing and storing powers of the permutations ρ_1 and ρ_2, implement a constant-time, table-driven colouring algorithm for perfect colourings represented as above.

    (b) As a function of k, the number of colours, what is the worst-case number of permutations that must be stored?

12. Write a program for interactive rendering and editing of isohedral tilings. It should be possible to pan, zoom and rotate a patch of tiles rendered within a rectangular window. There should be a separate window containing a single prototile whose shape and tiling vertex parameterization can be edited. Changes to the prototile shape should be reflected immediately in the rendered tiling. The program should also allow the patch of tiles to be exported to a vector drawing.

This programming task represents a significant effort. Here is a suggestion for a sequence of steps to follow in constructing such a program.

(a) Implement a single tiling type with simple 4^4 topology, fixed tiling vertices, and a single non-trivial edge shape. Good choices are IH61, IH62, IH71 and IH73.

(b) Extend the implementation to the other isohedral types with 4^4 topology but fixed square tiling vertices (i.e., those with tiling vertex parameterization TV35).

(c) Add support for tiling types with vertex parameterization TV30 (rectangles).

(d) Continue until the implementation can support all isohedral types of topological type 4^4.

(e) Finally, extend to the other topological types.

13. Write a program to create parquet deformations: patches of tiles that slowly evolve between two "keyframe" tile shapes placed at the left and right extremes of the patch. Parquet deformations were devised by Huff, and later popularized by Hofstadter [27, Chapter 10]. My Bridges 2008 paper (available at `http://www.cgl.uwaterloo.ca/~csk/papers/kaplan_bridges2008.pdf`) discusses methods for drawing parquet deformations based on isohedral tilings [32]. Restrict your attention to the isohedral types mentioned in part (b) of the previous question.

14. Section 5.4.1 shows how the tiling vertex parameterization for an isohedral type may be thought of as the product of a matrix of coefficients with a vector of parameters. The result is a convenient, table-driven means of representing the vertex parameterizations. Can a similar procedure be applied to the aspect transforms and translation vectors?

Choose an isohedral type with at least two aspects and at least two tiling vertex parameters. Derive the entries of the aspect transformation matrices and the coordinates of the translation vectors as expressions in terms of the tiling vertex parameters.

CHAPTER 6

Nonperiodic and Aperiodic Tilings

Aperiodic tilings have received a great deal of attention over the past few decades, both from tiling theorists and lay audiences. Perhaps most well known are those invented by Penrose. Two examples are shown in Figure 6.1, one constructed from "kites" and "darts" and the other from rhombs of two sizes.

A tiling that is not periodic is called *nonperiodic*. A frequent but incorrect assumption is that the notions of aperiodicity and nonperiodicity coincide. In fact, the aperiodic tilings are a very special subset of the nonperiodic tilings. It is worth clarifying the distinction between the two (which is similar to the distinction made in Chapter 5 between k-isohedral and k-anisohedral), showing why the aperiodic tilings are an active and exciting area of research.

Lack of periodicity is not in itself a very surprising property. It is easy to construct tilings where every tile shape is unique—consider, for example, the Voronoi diagram induced by an infinite integer lattice whose points have been jittered randomly. Clearly there can be no hope of periodicity in such a

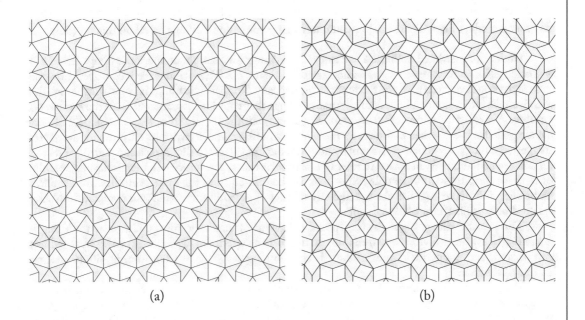

(a)   (b)

Figure 6.1: The two famous aperiodic tilings of Penrose. The "kite and dart" tiling is shown in (a) and thin and thick rhombs in (b).

case, but this fact seems unimpressive. In developing a definition of aperiodicity, we therefore consider only those nonperiodic tilings with a finite number of prototiles, i.e., those that are k-hedral for some k.

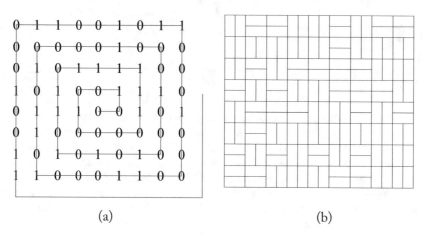

(a)   (b)

Figure 6.2: A contrived example of how even a very simple shape may yield nonperiodic tilings. A spiral path is used to place digits from the binary expansion of π. Each digit is then used to place a pair of bricks, oriented vertically to represent a 0 and horizontally to represent a 1. The resulting tiling, when extended to the whole plane, is (probably) nonperiodic, even though the brick prototile could easily be used to construct periodic tilings. There exist uncountably many nonperiodic tilings based on this prototile.

Under this restricted definition, we can still construct very simple nonperiodic tilings. Even a 2-by-1 brick yields an infinite variety, as demonstrated in Figure 6.2. These tilings seem contrived, however, because the same brick can easily be made to tile periodically. Nonperiodic tilings become truly interesting when we take into account all possible alternative tilings that could be constructed from the same set of prototiles. We call a set of prototiles an *aperiodic tile set* when the set admits at least one tiling, but none that are periodic. We can then define an *aperiodic tiling* as a tiling whose prototiles are an aperiodic tile set. An aperiodic tiling is one that is "essentially nonperiodic", in the sense that no rearrangement of its tiles will achieve periodicity. When used to refer to a particular tiling, aperiodicity is, therefore, a far reaching concept—it encompasses all possible tilings that can be formed from the same prototiles. As a result, it can be very difficult to establish the aperiodicity of a set of prototiles. Periodicity is comparatively easy: one need only exhibit a translational unit.

The Penrose tilings highlight the special behaviour of aperiodic tilings. Consider the tiling in Figure 6.1(b). By themselves, the two rhombs do not form an aperiodic tile set; they can be arranged into both periodic and nonperiodic tilings. There is also nothing stopping us from forming tilings that use just one of the rhombs.

When we wish to restrict the ways that prototiles may fit together, a common practice is to augment them with *matching conditions*. Matching conditions typically take the form of symbolic or geometric constraints on tiling edges (or sometimes vertices) that must be satisfied in any tiling. For example, prototile edges might be given labels, together with a rule indicating which pairs of labels are compatible. We then ask that in any tiling by these prototiles, two tiles that share an edge have compatible labels on that edge. Most simple matching conditions can also be encoded as indentations

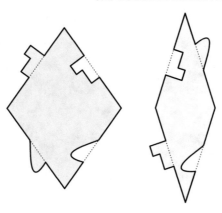

Figure 6.3: Sample matching conditions on the rhombs of Penrose's aperiodic tile set $P3$. The unmodified rhombs (indicated by dotted lines) can form many periodic tilings. The puzzle-piece deformations on the tile edges guarantee that any tiling formed from these new shapes will be nonperiodic.

and protrusions on prototile edges; tiles are then forced to snap together like a jigsaw puzzle. A set of protrusions that express the matching conditions is shown in Figure 6.3. It is these two modified shapes that form an aperiodic tile set, known as the Penrose tile set $P3$ (the modified kite and dart are known as $P2$). Many sets of prototiles must be endowed with similar matching conditions to enforce aperiodicity. The matching conditions are typically not shown when the tilings are rendered, perhaps leading to the confusion between nonperiodicity and aperiodicity.

Matching conditions are more important in mathematical proofs than in algorithmic applications of tilings. Many interesting tilings like the Penrose tilings come equipped with straightforward tiling algorithms that generate patches of tiles known to obey the matching rules. The rules are more important in proving that the only tilings admitted by the same prototiles are those intended by the algorithm.

In computer graphics applications, true aperiodicity is rarely a requirement. We often ask only that we can construct large patches of tiles that do not appear too "orderly". Many algorithms might produce such patches, as part of tilings that are nonperiodic without being aperiodic. This chapter will explore both nonperiodicity and aperiodicity, returning to Penrose tilings after a discussion of substitution systems and rep-tiles, and an introduction to Wang tiles.

6.1   SUBSTITUTION TILINGS AND REP-TILES

A square can trivially be divided into four smaller congruent squares. We can then scale the subdivided figure by a factor of two to obtain four squares congruent to the original. By iterating this procedure, we can build ever larger patches of squares. Without any appeal to periodicity, The Extension Theorem (Section 2.4) tells us that we can, in fact, tile the plane with squares. Alternatively, we can cover any region of the plane with squares to any level of subdivision by surrounding the region with a single large square and subdividing repeatedly, omitting the rescaling operation.

It is natural to expect that a similar process would hold any time we can define a set of prototiles, each of which is expressible as a patch of scaled copies of shapes from the same set. Consider, for example, the kite and domino prototiles in Figure 6.4. Each is congruent to a union of scaled kites and dominoes.

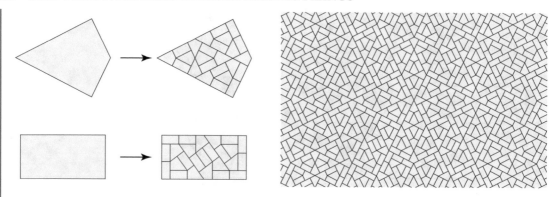

Figure 6.4: A simple example of a substitution tiling, the Kite-Domino tiling by Frettlöh and Baake. The prototiles are made from edges of side lengths 1 and 2. The diagram on the left shows how each of the prototiles is composed from smaller tiles. A sample patch of tiles appears on the right.

By repeatedly replacing individual tiles by these unions, we can construct the tiling shown in the figure. This style of replacement is very general; we require only that the same scaling factor be used consistently everywhere.

This simple example captures many of the fundamental properties of *substitution tilings*. Generally, a *substitution system* consists of three elements:

- A prototile set $\{T_1, \ldots, T_n\}$;

- A scaling factor s, known as the *expansion constant*; and

- A set of n rules that define, for each i, a patch of tiles from the prototile set that is congruent to a copy of T_i scaled by s.

A *substitution tiling* is then defined to be a tiling in which every patch is a sub-patch of one created by starting with a single prototile and applying the substitution rules a finite number of times. (Under this definition, every tiling created by applying the substitution rules is immediately a substitution tiling. It also allows a slightly larger set of possibilities that do not follow directly from the rules; these possibilities are of mathematical interest, but do not affect the discussion here.)

The definition of a substitution system enforces the consistent scaling requirement mentioned above by establishing a single expansion constant s that is used across all substitution rules. A substitution rule can be written in the form $T_i \rightarrow \{(j_1, M_1), \ldots, (j_{r_k}, M_{r_k})\}$. Each pair consists of an index between 1 and n, together with a rigid motion. The rule defines a patch consisting of a copy of T_{j_1} transformed by M_1, a copy of T_{j_2} transformed by M_2, and so on. In a substitution tiling, this patch will be congruent to a copy of T_i scaled by s. Note that we can also compose all of the rigid motions in the rules with a uniform scaling by $1/s$, in which case the rules will produce patches congruent to the original prototiles. Figure 6.5 shows a complete example of a substitution tiling (discovered by Ammann) that includes the coordinates of the prototile vertices and substitution rules expressed in this manner.

Once an appropriate substitution system has been defined, it is easy to apply the substitution rules to produce a patch of tiles of any size. Given a region R to be covered with tiles, the simplest approach is

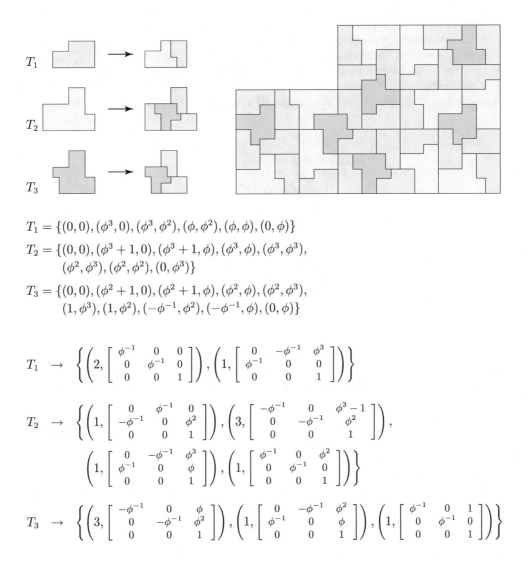

$T_1 = \{(0,0), (\phi^3, 0), (\phi^3, \phi^2), (\phi, \phi^2), (\phi, \phi), (0, \phi)\}$

$T_2 = \{(0,0), (\phi^3 + 1, 0), (\phi^3 + 1, \phi), (\phi^3, \phi), (\phi^3, \phi^3),$
$\quad\quad (\phi^2, \phi^3), (\phi^2, \phi^2), (0, \phi^3)\}$

$T_3 = \{(0,0), (\phi^2 + 1, 0), (\phi^2 + 1, \phi), (\phi^2, \phi), (\phi^2, \phi^3),$
$\quad\quad (1, \phi^3), (1, \phi^2), (-\phi^{-1}, \phi^2), (-\phi^{-1}, \phi), (0, \phi)\}$

$$T_1 \;\rightarrow\; \left\{ \left(2, \begin{bmatrix} \phi^{-1} & 0 & 0 \\ 0 & \phi^{-1} & 0 \\ 0 & 0 & 1 \end{bmatrix}\right), \left(1, \begin{bmatrix} 0 & -\phi^{-1} & \phi^3 \\ \phi^{-1} & 0 & 0 \\ 0 & 0 & 1 \end{bmatrix}\right) \right\}$$

$$T_2 \;\rightarrow\; \left\{ \left(1, \begin{bmatrix} 0 & \phi^{-1} & 0 \\ -\phi^{-1} & 0 & \phi^2 \\ 0 & 0 & 1 \end{bmatrix}\right), \left(3, \begin{bmatrix} -\phi^{-1} & 0 & \phi^3 - 1 \\ 0 & -\phi^{-1} & \phi^2 \\ 0 & 0 & 1 \end{bmatrix}\right), \right.$$
$$\left. \left(1, \begin{bmatrix} 0 & -\phi^{-1} & \phi^3 \\ \phi^{-1} & 0 & \phi \\ 0 & 0 & 1 \end{bmatrix}\right), \left(1, \begin{bmatrix} \phi^{-1} & 0 & \phi^2 \\ 0 & \phi^{-1} & 0 \\ 0 & 0 & 1 \end{bmatrix}\right) \right\}$$

$$T_3 \;\rightarrow\; \left\{ \left(3, \begin{bmatrix} -\phi^{-1} & 0 & \phi^2 \\ 0 & -\phi^{-1} & \phi^2 \\ 0 & 0 & 1 \end{bmatrix}\right), \left(1, \begin{bmatrix} 0 & -\phi^{-1} & \phi^2 \\ \phi^{-1} & 0 & \phi \\ 0 & 0 & 1 \end{bmatrix}\right), \left(1, \begin{bmatrix} \phi^{-1} & 0 & 1 \\ 0 & \phi^{-1} & 0 \\ 0 & 0 & 1 \end{bmatrix}\right) \right\}$$

Figure 6.5: A complete visual and symbolic description of the substitution system for Ammann's aperiodic tile set $A3$. The substitution rules are shown pictorially at the top, together with a patch of tiles produced from T_1 after four rounds of substitution. The first three equations give coordinates for the shape vertices of the three prototiles. These equations are followed by full substitution rules, as explained in Section 6.1. The matrices define 2D affine transformations in homogeneous coordinates. All coordinates and rules are expressed in terms of the golden ratio $\phi = (1 + \sqrt{5})/2$, which is also the tiling's expansion constant.

to choose an initial prototile T_i and transform it so that it surrounds R. We then apply the substitution rules, starting with this single prototile, as many times as desired (we incorporate a scaling by $1/s$ into each rule's transformations, as suggested above). A simple recursive implementation that accumulates transformations, similar to the rendering of self-similar curves like the Koch snowflake, can serve as a drawing algorithm. The following pseudocode defines a procedure DRAW that consumes an index i_0 in an array of prototiles, an affine transformation M_0, a recursive depth l, and an array of substitution rules. It relies on a helper procedure DRAWONETILE(i, M) that draws a copy of prototile T_i transformed by M.

```
procedure DRAW(i₀, M₀, l, rules):
    if l = 0 do
        DRAWONETILE(i₀, M₀)
    else do
        rule ← rules[i₀]
        for each (i, M) in rule do
            DRAW(i, M₀M, l − 1, rules)
        end for
    end if
end procedure
```

Unlike the examples in Figures 6.4 and 6.5, a substitution rule need not replace a tile T_i with a patch of smaller tiles whose union is congruent to T_i. The substituted tiles can extend beyond the parent or leave indentations, as long as those inconsistencies are accounted for by corresponding geometry in neighbouring subdivided tiles. An example of a substitution system of this form is given in Figure 6.6. When the substitution rules are "exact", they are referred to as *compositions* by Grünbaum and Shephard [24, Section 10.1]. The resulting tilings are also sometimes called *volume hierarchic*. Frank takes the opposite approach, using the term *pseudo-self-similar* for tilings in which the rules do not produce patches that are congruent to scaled prototiles [16].

When describing substitution systems, it is occasionally useful to allow two or more prototiles to be congruent. Figure 6.6 shows an example. In that tiling, there are two distinct triangular prototiles, labelled 1 and 2. These triangles must be kept distinct; they require different rules in order to produce a consistent tiling under repeated substitution. Congruent prototiles can be considered distinct based on their labels, or they can be merged together after substitution is complete. Of course, the implementation suggested above does not change in the presence of congruent prototiles.

Substitution rules may also produce tiles that overlap after multiple substitution steps. Overlaps are permissible as long as they are consistent: the two tiles that overlap must be copies of the same prototile, transformed by the same similarity. Figure 6.7 shows a substitution tiling in which the rules give rise to overlaps.

When there are congruent prototiles, a visualization of the substitution rules must show the full identities of the tiles and not just their shapes. In Figure 6.6, original and substituted tiles are shown with their numeric labels. Furthermore, when prototiles are symmetric, we must distinguish between several possible similarity transformations that might position a copy of the tile in a substitution. The distinction can be visualized with a suitable marking on the tiling. The labels are used as markings in Figure 6.6, and triangles are used in Figure 6.7.

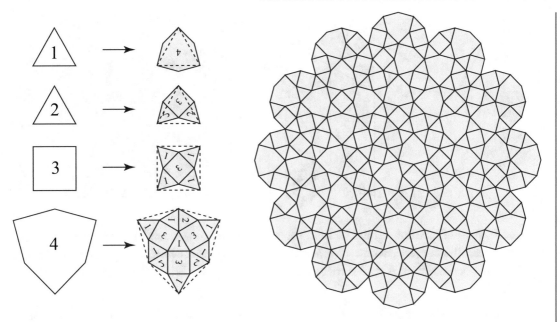

Figure 6.6: A set of substitution rules for Gähler's Shield tiling, together with a sample patch of tiles. The substituted tiles leave indentations and protrusions in their parents' outlines. Each prototile is marked with a number; the numbers are shown transformed on the right-hand sides of the rules to indicate which prototiles should be used and how they should be transformed.

In the special case where $n = 1$, the single prototile is called a *rep-tile*. Specifically, a k-rep-tile is a shape that can be composed from k scaled copies of itself, all congruent to each other. All triangles and parallelograms are 4-rep-tiles. Two well-known, non-trivial examples are the Chair tiling and the Sphinx tiling; both are shown in Figure 6.8.

The two rep-tiles mentioned above produce nonperiodic tilings. Indeed, many substitution systems can be shown to produce nonperiodic tilings, provided the rules satisfy a fairly weak uniqueness criterion [24, Theorem 10.1.1]. Of course, these tilings might not be aperiodic; as with the Penrose rhombs of Figure 6.1, it might be possible to find a periodic arrangement of the same prototiles. However, this deficiency can be remedied; Goodman-Strauss has shown that subject to a very mild condition, the tiles in a substitution tiling can be decorated with matching conditions in a way the enforces the structure generated by the substitution rules [20]. In this way, most substitution systems give rise to aperiodic tilings.

Polyominoes [19] can serve as a rich source of substitution tilings and rep-tiles. A polyomino is a finite set of squares, joined edge-to-edge. Many polyominoes are rep-tiles and lead immediately to tilings of the plane. More generally, if a set of polyominoes can be assembled into any rectangle, then they can serve as prototiles in a substitution system. We arrange sufficiently many of these rectangles to form a square, and substitute this square for every unit in each of the prototiles.

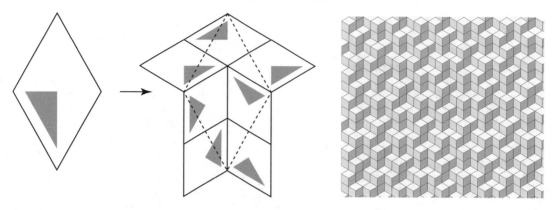

Figure 6.7: A set of substitution rules for Lord's nonperiodic tiling, together with a sample patch of tiles. The rules cause tiles to overlap with each other, but in such a way that overlaps happen in their entirety, avoiding any inconsistencies. A shaded triangle is used to indicate the orientations of substituted tiles.

There are still many unsolved problems related to substitution tilings and rep-tiles, and in general it can be difficult to design a substitution system that produces a legal tiling. However, once such a system is devised, rendering (patches of) tilings algorithmically is straightforward.

A large collection of substitution tilings (including those illustrated in Figures 6.4–6.8) is maintained by Harriss and Frettlöh at `http://tilings.math.uni-bielefeld.de/`.

6.2   WANG TILES AND APERIODICITY

In Section 2.4, I mentioned the *Tiling Problem*: does a given set of shapes admit any tilings of the plane? A solution to the tiling problem should take the form of an algorithm that decides in finite time whether or not any tilings exist.

Let us examine this question in the context of a restricted family of tiles. *Wang tiles* are square prototiles with marked edges. The markings are usually indicated by colours, with the understanding that like colours must meet across edges in any tiling by these tiles. We impose two additional restrictions. First, tiles must meet edge-to-edge. Second, they must be placed in a tiling by translation only—rotations and reflections are forbidden. Note that both of these restrictions can be expressed geometrically if desired. Wang tiles effectively reduce the problem of fitting tiles together to one that is discrete and combinatorial.

As explained by Grünbaum and Shephard, Hao Wang articulated four possible outcomes for any set of shapes [24, Section 11.3]:

1. They do not admit any tilings.

2. They admit tilings that are always periodic.

3. They admit both periodic and nonperiodic tilings.

4. They admit tilings that are always nonperiodic.

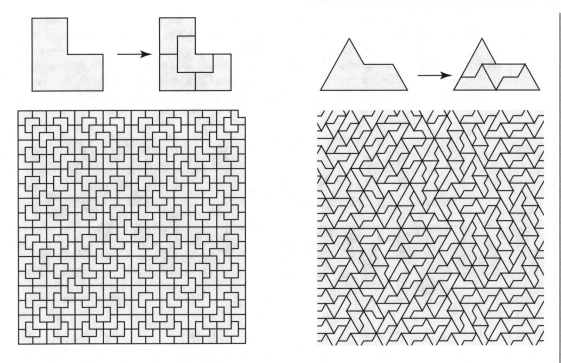

Figure 6.8: Two well-known examples of rep-tiles: the Chair tiling on the left and the Sphinx tiling on the right.

The fourth case is precisely our definition of an aperiodic tile set. In 1961, Wang conjectured that this situation could never arise—a reasonable conjecture at the time, but one that we now know to be false. With this assumption in hand, he was able to formulate an algorithm that would decide the Tiling Problem for marked square tiles. The algorithm iterates over the positive integers. For each integer m, it constructs all possible $m \times m$ blocks of prototiles. If the given Wang tiles admit any periodic tilings, the algorithm will eventually find an m for which there exists an $m \times m$ translational unit. If not (and assuming that Case 4 is impossible), the algorithm will encounter an m for which no $m \times m$ block can be constructed that is consistent with the tile markings (the Extension Theorem guarantees the existence of such an m).

In 1966, Berger presented the first aperiodic tile set, a collection of over 20 000 Wang tiles that admit only nonperiodic tilings [2]. Berger's discovery invalidates Wang's algorithm because the algorithm assumes that a set of prototiles will tile via a finite translational unit or not at all. It was subsequently shown that Wang tiles could be used to simulate Turing machines [18], with a set of prototiles tiling the plane if and only if a corresponding Turing machine never halts. The halting problem is, therefore, reducible to the tiling problem, proving that the latter must be undecidable.

Since Berger's initial discovery, tiling theorists have sought smaller aperiodic sets of Wang tiles. The current record holder is a set of 13 prototiles [8]. Of course, smaller aperiodic sets exist outside the restrictions of Wang tiles. In 1971 Robinson demonstrated an aperiodic set of six prototiles, which offer some intriguing design possibilities [24, Section 10.2]. Other small aperiodic tile sets have been discovered by Ammann [24, Section 10.4] and Goodman-Strauss [21]. The two-tile sets by Penrose will be discussed in the next section. However, nobody has been able to improve upon these results or to show that at least two tile shapes are required. The question of whether there exists an *aperiodic monotile*, a single shape that tiles only aperiodically, remains open [7, Section C18]. Almost nothing is known about what must or must not be true about such a shape, and the problem remains one of the most beautiful in geometry.

Figure 6.9: An "unrestricted" set of Wang tiles with two edge colours. For any choices of colours to the north and west of a tile location, there are always two possible tiles that can be placed compatibly in that location.

In computer graphics, Wang tiles are gaining popularity as a way to cover the plane with tiles while avoiding obvious repetition [5, 35]. In this context, aperiodicity is irrelevant; what matters is the ability to produce nonperiodic tilings efficiently. For this purpose, a set of eight Wang tiles can be described over two colours (or four, if we distinguish between horizontal and vertical occurrences of the same colour), as shown in Figure 6.9. This set has the property that given a tile location with any two tiles adjacent across the northern and western borders, there are always two tiles that can be placed in that location without violating the matching conditions. We can therefore construct an arbitrarily large patch of tiles without obvious repetitions simply by filling the region in row-major order, always choosing randomly from the two possibilities available at every step. The key in graphics applications is to fill the tiles with information (textures, sample positions, etc.) that embodies the underlying matching rules. Since the contents of the tiles are usually computed once ahead of time, it is relatively easy to suppress artifacts further by adding more tiles. Note that this technique also adapts naturally to three dimensions.

Much more information about the uses of Wang tiles in computer graphics can be found in the Synthesis Lecture by Lagae [36].

6.3   PENROSE TILINGS

As was mentioned at the beginning of this section, the Penrose tile sets $P2$ (the "kite" and "dart") and $P3$ (thin and thick rhombs) are both aperiodic with suitable matching conditions (illustrated for $P3$ in Figure 6.3). Of course, proving the aperiodicity of $P2$ or $P3$ is difficult. First, one must establish that the matching conditions prevent the prototile set from admitting any periodic tilings. Then one must show

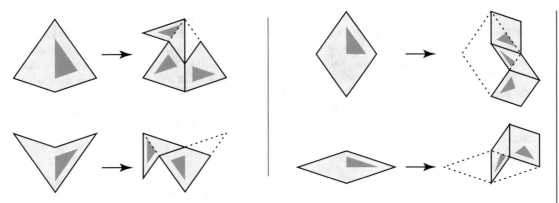

Figure 6.10: A set of substitution rules for Penrose's aperiodic tile sets $P2$ and $P3$. These rules are often presented in the style of the Lord tiling in Figure 6.7 where overlapping tiles must be pruned after substitution. Here, the rules fill the plane exactly, without introducing any overlaps. Shaded triangles are used to indicate the orientations of substituted tiles.

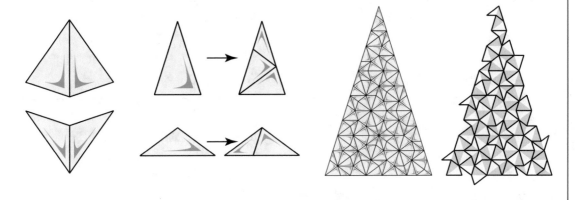

Figure 6.11: Modified substitution rules for Penrose's set $P2$. The kite and dart are each bisected into direct and reflected copies of isosceles triangles (orientation is indicated by a shaded marking). The substitutions for both triangles are shown. Repeated substitution exactly fills the initial patch of tiles. On the far right, the final configuration of kites and darts can be determined consistently from one of the orientations of each of the two triangles.

that it *does* admit at least one nonperiodic tiling. Details of the proof can be found in Grünbaum and Shephard [24, Section 10.3].

Of course, for computer graphics purposes, we are primarily concerned with the problem of filling arbitrary regions of the plane with interlocking tiles from either set; the underlying logic is only of academic interest. There are several very different algorithms for laying out Penrose tilings. Some of these algorithms are themselves sophisticated results in algebra, geometry, and number theory. Details on the so-called "multigrid" and "lattice projection" methods can be found in the book by Senechal [46].

On the other hand, a rendering approach based on substitution leads to a far simpler implementation. Figure 6.10 shows substitution rules for $P2$ and $P3$, which can be applied recursively to draw finite patches of any size. These rules are very efficient in that they never lead to overlapping tiles, which would then need to be detected and coalesced. On the other hand, they are less than ideal for filling regions of the plane with tiles. The substituted tiles do not entirely cover their parents, and so it does not suffice simply to ensure that a given region is completely surrounded by a single tile—substitution might ultimately leave part of the region uncovered.

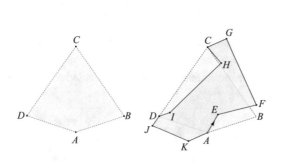

$$A = (0, 0)$$
$$B = (\cos(\tfrac{\pi}{10}), \sin(\tfrac{\pi}{10}))$$
$$C = (0, \tfrac{1+\sqrt{5}}{2})$$
$$D = (-\cos\tfrac{\pi}{10}, \sin\tfrac{\pi}{10})$$

$$E = A + r_1(\cos(\theta_1 + \tfrac{\pi}{10}), \sin(\theta_1 + \tfrac{\pi}{10}))$$
$$F = B + r_2(\theta_2 + \cos(\tfrac{7\pi}{10}), \sin(\theta_2 + \tfrac{7\pi}{10}))$$
$$G = C + r_1(\cos(\theta_1 - \tfrac{\pi}{10}), \sin(\theta_1 - \tfrac{\pi}{10}))$$
$$H = \text{rotate}(C, -\tfrac{2\pi}{5}, H)$$
$$I = \text{rotate}(C, -\tfrac{2\pi}{5}, G)$$
$$J = \text{rotate}(A, -\tfrac{4\pi}{5}, G)$$
$$K = \text{rotate}(A, -\tfrac{4\pi}{5}, F)$$

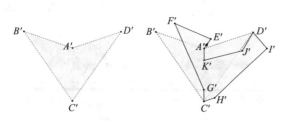

$$A' = (0, 0)$$
$$B' = (-\cos\tfrac{\pi}{10}, \sin\tfrac{\pi}{10})$$
$$C' = (0, -1)$$
$$D' = (\cos(\tfrac{\pi}{10}), \sin(\tfrac{\pi}{10}))$$

$$E' = A' + r_2(\cos(\theta_2 + \tfrac{\pi}{2}), \sin(\theta_2 + \tfrac{\pi}{2}))$$
$$F' = B' + r_1(\cos(\theta_1 - \tfrac{\pi}{10}), \sin(\theta_1 - \tfrac{\pi}{10}))$$
$$G' = C' + r_2(\cos(\theta_2 + \tfrac{7\pi}{10}), \sin(\theta_2 + \tfrac{7\pi}{10}))$$
$$H' = \text{rotate}(C', -\tfrac{2\pi}{5}, G')$$
$$I' = \text{rotate}(C', -\tfrac{2\pi}{5}, F')$$
$$J' = \text{rotate}(A', -\tfrac{4\pi}{5}, F')$$
$$K' = \text{rotate}(A', -\tfrac{4\pi}{5}, E')$$

Figure 6.12: A tiling vertex parameterization for generalized Penrose kites and darts, controlled by four real-valued parameters $r_1, \theta_1, r_2,$ and θ_2. The vertices are enumerated in counterclockwise order starting at A for the kite and A' for the dart. The function rotate(p, θ, q) rotates point q by angle θ about point p.

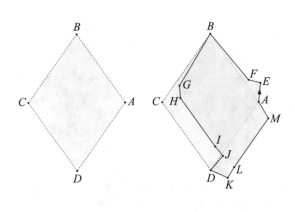

$A = (\sin\frac{\pi}{5}, 0)$
$B = (0, \cos\frac{\pi}{5})$
$C = (-\sin\frac{\pi}{5}, 0)$
$D = (0, -\cos\frac{\pi}{5})$

$E = A + r_1(\cos(\frac{7\pi}{10} - \theta_1), \sin(\frac{7\pi}{10} - \theta_1))$
$F = A + r_2(\cos(\frac{7\pi}{10} - \theta_2), \sin(\frac{7\pi}{10} - \theta_2))$
$G = \text{rotate}(B, -\frac{2\pi}{5}, F)$
$H = \text{rotate}(B, -\frac{2\pi}{5}, E)$
$I = D + r_2(\cos(\frac{\pi}{2} - \theta_2), \sin(\frac{\pi}{2} - \theta_2))$
$J = D + r_1(\cos(\frac{\pi}{2} - \theta_1), \sin(\frac{\pi}{2} - \theta_1))$
$K = \text{rotate}(D, -\frac{2\pi}{5}, J)$
$L = \text{rotate}(D, -\frac{2\pi}{5}, I)$
$M = \text{rotate}(D, -\frac{2\pi}{5}, H)$

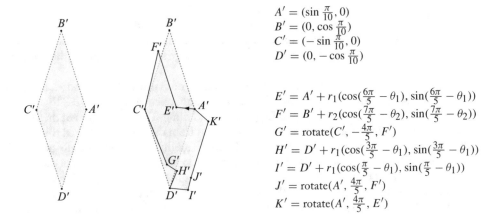

$A' = (\sin\frac{\pi}{10}, 0)$
$B' = (0, \cos\frac{\pi}{10})$
$C' = (-\sin\frac{\pi}{10}, 0)$
$D' = (0, -\cos\frac{\pi}{10})$

$E' = A' + r_1(\cos(\frac{6\pi}{5} - \theta_1), \sin(\frac{6\pi}{5} - \theta_1))$
$F' = B' + r_2(\cos(\frac{7\pi}{5} - \theta_2), \sin(\frac{7\pi}{5} - \theta_2))$
$G' = \text{rotate}(C', -\frac{4\pi}{5}, F')$
$H' = D' + r_1(\cos(\frac{3\pi}{5} - \theta_1), \sin(\frac{3\pi}{5} - \theta_1))$
$I' = D' + r_1(\cos(\frac{\pi}{5} - \theta_1), \sin(\frac{\pi}{5} - \theta_1))$
$J' = \text{rotate}(A', \frac{4\pi}{5}, F')$
$K' = \text{rotate}(A', \frac{4\pi}{5}, E')$

Figure 6.13: A tiling vertex parameterization for generalized Penrose rhombs, controlled by four real-valued parameters $r_1, \theta_1, r_2,$ and θ_2. The vertices are enumerated in counterclockwise order starting at A for the thick rhomb and A' for the thin rhomb.

To obtain more control over the substitution process, we can use modified rules based on an analysis by Robinson, as described by Grünbaum and Shephard. Figure 6.11 illustrates the process for *P*2 (a similar argument applies to *P*3). Each prototile can be divided into left-handed and right-handed isosceles triangles by splitting them along lines of mirror reflection. These triangles can be given their own substitution rules. After any desired number of substitutions, the right-handed triangles can be discarded and the kites and darts can be recovered consistently from the left-handed triangles alone. The resulting patch will still fail to cover the entire starting tile exactly, but the indentations can be worked around easily.

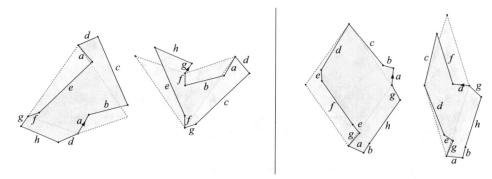

Figure 6.14: Edge labels for the tiling edges of the two sets of Penrose tiles, in the spirit of the incidence symbols used for the isohedral tilings. The kite and dart are shown on the left, and the two rhombs on the right. (The edge labels are not related between the two sets.) Pairs of labels correspond as described in the text.

What about the shapes of the tiles themselves? The substitution systems above can provide us with a list of locations at which to draw prototiles. Geometric matching conditions such as those shown in Figure 6.3 suggest that we are free to choose two arbitrary ∫ edge shapes to make up the outlines of the tiles.

Unfortunately, this interpretation of the possible shapes of Penrose tiles is limited, as can be seen in Grünbaum and Shephard's reproduction of Penrose's aperiodic chicken tiling [24, Figure 10.3.13]. They overlay the chickens with the corresponding unmodified tiling. The registration of these two tilings reveals that the chickens have tiling vertices that are different from those of the original tiling! Although these tiling vertices can be simulated by introducing degeneracies in the form of partially overlapping tile edges, there is some benefit in being able to parameterize the locations of these vertices directly.

However, we cannot simply provide a tiling vertex parameterization as was done in Section 5.3 for the isohedral tiling types. In a tiling by $P2$ or $P3$, copies of a given prototile will be surrounded in a finite number of different ways, causing tiling vertices to be located differently around its boundary. But it is possible to parameterize an extended set of *quasivertices*, points on a prototile's boundary that are tiling vertices anywhere in a tiling, or that are forced into existence by those tiling vertices. Quasivertex parameterizations are provided for $P2$ in Figure 6.12 and for $P3$ in Figure 6.13.

Given this extended set of quasivertices, we must revise the original matching conditions to account for the new tile edges that have been introduced. Taking inspiration from the use of incidence symbols in isohedral tilings, the possible edge shapes can be specified by labeling the edges around each tile and indicating adjacency rules for the labels. The edges of the kite and dart can be labeled *abcdaefghd* and *ghefgcdabf*, respectively, where the enumerations start at the edges marked with arrows in Figure 6.12. To enforce matching between adjacent tiles, we require that the pairs (a, d), (b, h), (c, e), and (f, g) interlock. In effect, a given kite and dart will have only four non-congruent edge shapes between them. Similarly, we can label the edges of the thick and thin rhombs, respectively, as *abcdefegabhg* and *afcdegabhg*, with the requirement that pairs (a, g), (b, e), (c, d), and (f, h) interlock. These labelings of the edges of the Penrose tiles are shown in Figure 6.14.

EXERCISES

1. Let S be the set of prototiles of the Archimedean tiling (4.8^2): a square and a regular octagon with the same edge length. As given, this prototile set cannot force the octagon to be used—S also admits the regular tiling (4^4). Modify the prototile shapes with geometric matching conditions that admit only the tiling analogous to (4.8^2) in such a way that:

   (a) All square tiles have the same orientation.

   (b) The square tiles come in two orientations.

2. Write a program that can construct patches based on a set of prototiles equipped with matching conditions on the edges. The prototiles are arbitrary polygons, to be placed edge-to-edge in the patch. Every polygon edge has an integer label. Two tiles that share an edge must have labels of the form i and $-i$ on that edge (and must also have the same edge length).

   The program should build the patch breadth-first. Begin with a single tile and place its edges in a queue. At every step, find a tile to place against the unused edge at the front of the queue, and place that tile's free edges at the back of the queue. Stop when the patch contains a predetermined number of tiles.

   There is no guarantee that the prototile set can generate a sufficiently large patch, and so a full implementation must be able to backtrack.

   A simple implementation should accept a set of prototiles with marked edges, together with the size of the patch to create (measured by the number of tiles it contains). It can draw the patch in a vector format (ideally with edge markings included) as output.

3. Consider the sequence of patches P_n, where P_1 contains a single kite from Penrose's aperiodic tile set $P2$, and each P_{k+1} is formed by applying the substitution rules from the left of Figure 6.10 to all tiles in P.

   (a) Let K_n and D_n represent the number of kites and darts in P_n. Determine the sequences K_n and D_n.

   (b) Use the sequences from part (a) to argue informally that the tiling formed by constructing P_n as n goes to infinity must be nonperiodic. (Hint: consider the ratio K_n/D_n).

4. Show that every triangle is a rep-tile.

5. (a) Show that the **P** pentomino shown on the right is a 4-rep-tile.

   (b) By tiling a square with copies of the **P** pentomino, show that it is also a 100-rep-tile.

6. Write a program that can render patches from substitution tilings, in the style of Section 6.1. Your program should accept an input file that contains a description of the shape of each prototile in a substitution set, together with the list of tile indices and similarities associated with each prototile's substitution rule. The program should also accept the index of a starting tile and a number of times to iterate the substitution process. Produce a vector drawing of the resulting patch as output. You can assume that the rules are provided in such a way that tiles will never overlap.

7. Modify the program of the previous question to handle substitution systems that might produce two coincident tiles (two copies of the same prototile that overlap in their entirety). A simple approach is to store, with each prototile, a list of the transformation matrices where that prototile appears in a patch, and to ensure that no matrix occurs more than once. However, care must be taken with symmetric prototiles—there will be multiple matrices that transform such a prototile to any given location.

8. Write a program that can construct large, rectangular patches of the unrestricted Wang tiles shown in Figure 6.9. Unlike that figure, do not render yellow and blue discs on the tiles—choose some other set of markings that uniquely identifies the tiles and exhibits visual continuity across tile edges. Better yet, allow the markings to be stored in an external file.

9. Write an interactive program that can render Lagae's variation on unrestricted Wang tiles. In his set, matching conditions are expressed with colours on the tile vertices; all tiles meeting at a vertex must share the same colour [37]. Instead of rendering a patch, render all tiles that intersect the viewing window, and support interactive panning and zooming. Use a hash function to select random colours for tiling vertices, allowing you to place tiles in the plane with random access.

10. Modify the general substitution renderer of Question 6 to handle the triangle-based substitution rules for Penrose's aperiodic set $P2$, as shown in Figure 6.11. You need to implement an extra post-processing step that discards half of the final triangles and renders kites and darts on top of those that remain.

11. Modify the Penrose renderer of the previous question to permit the edge shapes and quasivertices to be edited interactively, using the parameterization discussed in Section 6.3. In addition to showing a patch of Penrose tiles, you should display a single, large copy of each prototile. Use sliders to control the parameters r_1, θ_1, r_2, and θ_2. The user should be able to click and drag on tile edges to edit them interactively. Consider also supporting curved paths as tiling edges.

12. Using the parameterized kite and dart of Figure 6.12, either directly or via the implementation described in the previous question, find values for the quasivertex parameters such that the dart takes the form of a thick Penrose rhomb, and the kite becomes a union of thick and thin rhombs.

CHAPTER 7

Survey

I conclude this book with a brief survey of some previous research and applications in computer graphics that make use of tiling theory.

7.1 DRAWING PERIODIC TILINGS

Software specifically geared towards the construction of tilings of the plane has been around for nearly thirty years. For the most part, these tools are based on the Heesch tilings, which can be seen as a precursor to the isohedral tilings in which internal symmetries of tiles are not recognized (see Section 5.5).

Chow had a successful FORTRAN program [3] that let the user input the portion of a tile's boundary that is "independent", i.e., not constrained to some other part of the boundary through a symmetry of the tiling. The program then filled in the remaining part of the tile and replicated it in the plane. Chow also discussed possible applications of his software in manufacturing.

For many years, Kevin Lee has offered a commercial software package called KaleidoMania! (originally called TesselMania!) that makes it easy to draw and decorate Escher-like tilings. His system is geared towards the use of tilings as a tool for mathematics education, and the most recent version of KaleidoMania! includes tutorials, games and puzzles designed for teaching concepts of geometry.

Tupper's Tess (http://www.peda.com/tess) has traditionally allowed the user to create drawings belonging to the frieze and wallpaper groups. More recent versions of Tess now support the Heesch tilings directly. Like KaleidoMania!, Tess is geared towards pedagogical use.

There exist many software tools for creating drawings based on symmetry groups. In many such tools, it is easy to draw tilings, but there is no explicit use of tiling theory and no tiling-like constraints are imposed on the user. The classic example of such a tool is Kali (http://www.geom.umn.edu/java/Kali/). Weeks's KaleidoTile (http://www.geometrygames.org/KaleidoTile/) supports more drawing styles and non-Euclidean symmetry groups. A popular commercial product is Artlandia's Symmetry Works (http://artlandia.com/products/SymmetryWorks/), a plug-in for Adobe Illustrator.

7.2 DRAWING NONPERIODIC TILINGS

Quasitiler (http://www.geom.umn.edu/apps/quasitiler/) was an application that made it possible to visualize a wide variety of nonperiodic tilings produced via the "lattice projection method". These included tilings by the Penrose rhombs, as well as generalizations to rotational orders other than five. Quasitiler was originally written for the NeXT computer, and was then given a CGI-based web interface. Unfortunately, the back end is no longer in operation, and so Quasitiler is unavailable. Murray has created a Java implementation as part of a collection of screensavers (http://screensavers.dev.java.net/).

I have created a simple Java applet for exploring the parameterization of edge shapes and tiling vertices for Penrose tilings, as described in Section 6.3. It is available at http://www.cgl.uwaterloo.ca/~csk/software/penrose/.

7.3   ESCHER-LIKE TILINGS

M.C. Escher had a lifelong fascination with tilings of the plane. He spent years filling a notebook with drawings of tilings by fish, birds, people, and dozens of other lifelike forms [45]. Escher has long been a source of inspiration for mathematicians and computer scientists; it is natural to ask whether computer graphics can assist in the creation of Escher-like tilings.

The software mentioned above for drawing tilings can be seen as implementing a "forward" process of experimentation similar to Escher's manual work. Starting from a simple tile shape with adjacency rules, the user can modify edges and the software will enforce constraints that preserve tileability. With skill, intuition, and luck, the user can eventually produce a tiling in the style of Escher's drawings.

Is it possible to design an "inverse" algorithm for tile design? In other words, given an arbitrary "goal shape", can an algorithm find a monohedral tiling of the plane by a prototile that resembles that shape? David Salesin and I explored this question in the context of the isohedral tilings [33]. We developed an "Escherization" algorithm, a continuous optimization that searched over the parameterizations of the shapes of isohedral prototiles as presented in Section 5.3. The objective function for the optimization realized the tile shape parameters as a polygon, and compared this polygon with the goal shape using an efficient L^2 distance metric [1]. The resulting algorithm could discover attractive Escher-like tilings from a variety of real-world goal shapes.

In later work, we adapted the Escherization algorithm to several varieties of dihedral tilings [34]. Escher produced many dihedral tilings by beginning with one of his monohedral systems and dividing the prototile into two pieces with a path connecting two points on its boundary. We applied the same technique to the isohedral tilings, augmenting the isohedral shape parameterization with parameters controlling the end points and shape of a "splitting path". Here, the objective function constructs the two prototile shapes, compares each with its respective goal shape, and returns the maximum of the two comparisons. We further showed that restricting this representation to the special case of Dress's "Heaven and Hell patterns" [11] made possible the construction of metamorphoses in the style of Escher's *Sky and Water*. We also applied the same optimization framework to build Escher-like tilings based on the shape parameterizations of the Penrose tilings in Section 6.3.

Escher also created a small number of carved wooden sculptures featuring spherical interpretations of his tessellations. Using these as a starting point, Yen and Séquin created an "Escher Sphere Construction Kit" [53], a system that allows the user to design ornamental spherical tilings much as one could create Euclidean tilings in the drawing tools above. Their software was based on Grünbaum and Shephard's classification of the isohedral tilings of the sphere [23], analogous to the planar classification discussed in Section 5.2. As an added feature, the tilings they create could be exported to rapid prototyping hardware and constructed as real artifacts.

Some of Escher's most well-known tessellations are his "Circle Limit" prints, based on hyperbolic tilings. Dunham has studied Escher's hyperbolic drawings for many years, and has published several papers with similar drawings [13, 12]. In my doctoral thesis, I showed how some repeating patterns in the plane might automatically be transferred to non-Euclidean geometry [30, Section 4.7]. Jos Leys has created a large number of hyperbolic interpretation of Euclidean Escher tessellations (http://www.josleys.com/show_gallery.php?galid=325). In a recent paper, von Gagern and Richter-Gebert demonstrate a sophisticated technique for conformally transferring Euclidean ornaments into the hyperbolic plane [50]. Their technique works for all the wallpaper groups except o.

7.4   SAMPLING

Recently, substitution tilings have emerged as a powerful technique for generating fast sampling patterns with blue noise properties. Typically, a single sample location or a set of sample locations is pre-computed for each tile. Given a probability distribution (such as an importance map), a sampling pattern is then generated by applying the substitution rules selectively to individual tiles where greater sample density is needed.

Ostromoukhov et al. used this approach with Penrose tiles, demonstrating their technique in an environment mapping application [42]. They selectively apply rules from a substitution system derived from the Penrose rhombs, and place a sample in each remaining tile. Precomputed perturbation vectors are used to displace the samples to improve their statistical properties, based on the local configuration around each tile. More recently, Ostromoukhov presented a new sampling method based on selective subdivision of polyominoes [41]. The polyominoes avoid frequency-domain artifacts that arose in the use of Penrose tiles.

Kopf et al. developed a sampling method based on recursive subdivision of Wang tiles [35]. They generate a set of progressive Poisson distributions over each tile, which can be used for fine-grained control of sample density. Coarser changes can be handled by subdividing tiles. They can make random choices when placing every Wang tile in a grid, suppressing obvious repetition.

For more information about tile-based sampling methods, see the Synthesis Lecture by Lagae [36].

7.5   TEXTURE GENERATION

Another popular use of Wang tiles is to create non-repeating arrangements of textures or geometry. In these applications, each tile is decorated with a small fragment of an overall pattern, and tiles are joined together to extend the pattern over a large region. Clearly, a single tile can be decorated with a unrolled toroidal design to create a periodic pattern. In these techniques, the challenge is to ensure that an entire collection of decorated tiles can meet smoothly across their boundaries in multiple distinct configurations. This smoothness is usually achieved by representing a matching condition as a fragment of a pattern that straddles a tiling edge. Each Wang tile is then constructed by assembling the four fragments associated with its edge colours and merging those fragments somehow in the tile's interior.

Stam was the first to consider the use of Wang tiles for texturing [49]. He used a recursive subdivision scheme: a set of 16 Wang tiles, each with a substitution rule that satisfied the matching conditions. A recursive algorithm was therefore necessary to generate patches of tiles.

Cohen et al. used the simpler set of eight Wang tiles shown in Figure 6.9 [5], recognizing that the ability to assemble tiles stochastically is more relevant for graphics applications than true aperiodicity. They introduced the scanline algorithm for placing tiles, in which each new tile is chosen from the possibilities that are compatible with previously placed neighbours to the north and west. They also articulated the *corner problem*: Wang tiles with simple matching conditions have no control over their diagonal neighbours, a property which can lead to artifacts near tiling vertices. They solve this problem by augmenting the matching conditions with additional colour information for tile vertices, effectively increasing the number of colours used in the matching conditions. The combination of edge colours and vertex colours forces them to generate many different tiles. They apply their technique to texture synthesis and primitive distribution. More recently, Fu and Leung showed how this approach could be adapted to a quad-based parameterization of an arbitrary surface [17].

Lagae and Dutré showed that a similar approach could be applied to Wang tiles in which matching conditions are expressed purely at the vertices rather than the edges [37]. In their technique, a tile's vertices are given colours; four tiles meeting around a tiling vertex must have the same colour at that vertex. In this scheme, it is easier to control the continuity of information across tile corners (as well as edges). They also show that Wang tiles can be looked up with "random access": the identity of any tile in the plane can be computed numerically without first laying out all tiles to the north and west of it. A hash function is used to compute vertex colours repeatably anywhere in the plane. The tile for a particular lattice square can be chosen based on the colours at its vertices.

APPENDIX A

The Isohedral Tiling Types

This appendix offers a listing of the 81 isohedral tiling types that can be realized with unmarked tiles (the remaining 12 marked types are less important in computer graphics applications, and are omitted). The name of each tiling type is shown, together with a sample tiling belonging to that type. To the right of the picture is a listing of the template for the tiling type, as described in Section 5.4.5. This information is sufficient to construct a complete interactive application for manipulating the shapes of unmarked isohedral tilings and rendering the results. An ASCII version of the same information is available online at http://www.cgl.uwaterloo.ca/~csk/projects/escherization/.

Note that in some cases, the incidence symbols are not identical to the ones given by Grünbaum and Shephard. Some tiling types had their edges relabelled for consistency, allowing several types to share a single tiling vertex parameterization. From an implementation point of view, the fact that a given incidence symbol is no longer lexicographically first among its equivalents is, of course, irrelevant.

IH1

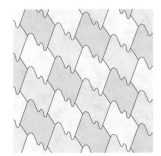

$3^6 \cdot [a^+b^+c^+d^+e^+f^+; d^+e^+f^+a^+b^+c^+]$
Parameterization TV1
Colouring (0) (1 2 0) (2 0 1)
Aspects 1
Rules
$\quad \vec{v_1} \leftarrow 1; 1$
$\quad \vec{v_2} \leftarrow 1; 2$

IH2

$3^6 \cdot [a^+b^+c^+d^+e^+f^+; b^-a^-f^+e^-d^-c^+]$
Parameterization TV2
Colouring (0 1) (1 2 0) (0 1 2)
Aspects 1, 1r
Rules
$\quad 1r \leftarrow 1; 1$
$\quad \vec{v_1} \leftarrow 1; 3$
$\quad \vec{v_2} \leftarrow 1r; 4$

IH3

$3^6 \cdot [a^+b^+c^+d^+e^+f^+; c^-e^+a^-f^-b^+d^-]$
Parameterization TV3
Colouring (0 1) (2 0 1) (2 0 1)
Aspects 1, 1r
Rules
$\quad$ 1r ← 1; 1
$\quad$ $\vec{v_1}$ ← 1; 2
$\quad$ $\vec{v_2}$ ← 1r; 6

IH4

$3^6 \cdot [a^+b^+c^+d^+e^+f^+; a^+e^+c^+d^+b^+f^+]$
Parameterization TV4
Colouring (0 1) (2 0 1) (2 0 1)
Aspects 1, 2
Rules
$\quad$ 2 ← 1; 1
$\quad$ $\vec{v_1}$ ← 1; 2
$\quad$ $\vec{v_2}$ ← 2; 4

IH5

$3^6 \cdot [a^+b^+c^+d^+e^+f^+; a^+e^+d^-c^-b^+f^+]$
Parameterization TV5
Colouring (0 1 2 1) (2 0 1) (2 0 1)
Aspects 1, 2, 1r, 2r
Rules
$\quad$ 2 ← 1; 1
$\quad$ 1r ← 1; 4
$\quad$ 2r ← 1r; 6
$\quad$ $\vec{v_1}$ ← 1; 2
$\quad$ $\vec{v_2}$ ← 1; 4, 6, 3, 1

IH6

$3^6 \cdot [a^+b^+c^+d^+e^+f^+; a^+e^-c^+f^-b^-d^-]$
Parameterization TV6
Colouring (0 1 2 2) (1 2 0) (0 1 2)
Aspects 1, 2, 1r, 2r
Rules
$\quad$ 2 ← 1; 1
$\quad$ 2r ← 2; 2
$\quad$ 1r ← 1; 2
$\quad$ $\vec{v_1}$ ← 1r; 2
$\quad$ $\vec{v_2}$ ← 2r; 6

IH7

$3^6 \cdot [a^+b^+c^+d^+e^+f^+; b^+a^+d^+c^+f^+e^+]$
Parameterization TV7
Colouring (0 1 2) (0 1 2) (0 1 2)
Aspects 1, 2, 3
Rules
$\quad 2 \leftarrow 1; 1$
$\quad 3 \leftarrow 1; 2$
$\quad \vec{v}_1 \leftarrow 2; 4$
$\quad \vec{v}_2 \leftarrow 3; 5$

IH8

$3^6 \cdot [a^+b^+c^+a^+b^+c^+; a^+b^+c^+]$
Parameterization TV1
Colouring (0) (1 2 0) (2 0 1)
Aspects 1
Rules
$\quad \vec{v}_1 \leftarrow 1; 1$
$\quad \vec{v}_2 \leftarrow 1; 2$

IH9

$3^6 \cdot [a^+b^+c^+a^+b^+c^+; a^+c^-b^-]$
Parameterization TV8
Colouring (0 1) (2 0 1) (0 1 2)
Aspects 1, 1r
Rules
$\quad 1r \leftarrow 1; 2$
$\quad \vec{v}_1 \leftarrow 1; 1$
$\quad \vec{v}_2 \leftarrow 1r; 5$

IH10

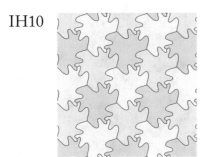

$3^6 \cdot [a^+b^+a^+b^+a^+b^+; b^+a^+]$
Parameterization TV9
Colouring (0) (1 2 0) (2 0 1)
Aspects 1
Rules
$\quad \vec{v}_1 \leftarrow 1; 1$
$\quad \vec{v}_2 \leftarrow 1; 6$

IH11

$3^6 \cdot [a^+a^+a^+a^+a^+a^+; a^+]$
Parameterization TV9
Colouring (0) (1 2 0) (2 0 1)
Aspects 1
Rules
$\quad \vec{v_1} \leftarrow 1; 1$
$\quad \vec{v_2} \leftarrow 1; 2$

IH12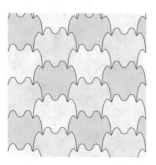

$3^6 \cdot [ab^+c^+dc^-b^-; dc^-b^-a]$
Parameterization TV10
Colouring (0) (1 2 0) (2 0 1)
Aspects 1
Rules
$\quad \vec{v_1} \leftarrow 1; 1$
$\quad \vec{v_2} \leftarrow 1; 2$

IH13

$3^6 \cdot [ab^+c^+dc^-b^-; db^+c^+a]$
Parameterization TV11
Colouring (0 1) (2 0 1) (0 1 2)
Aspects 1, 1r
Rules
$\quad 1r \leftarrow 1; 2$
$\quad \vec{v_1} \leftarrow 1; 1$
$\quad \vec{v_2} \leftarrow 1r; 6$

IH14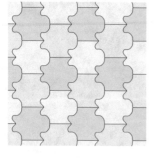

$3^6 \cdot [a^+b^+c^+c^-b^-a^-; c^-b^-a^-]$
Parameterization TV12
Colouring (0) (1 2 0) (2 0 1)
Aspects 1
Rules
$\quad \vec{v_1} \leftarrow 1; 1$
$\quad \vec{v_2} \leftarrow 1; 6$

IH15

$3^6 \cdot [a^+b^+c^+c^-b^-a^-; a^+b^-c^+]$
Parameterization TV13
Colouring (0 1) (2 0 1) (0 1 2)
Aspects 1, 2
Rules
   $2 \leftarrow 1; 1$
   $\vec{v_1} \leftarrow 1; 2$
   $\vec{v_2} \leftarrow 2; 3$

IH16

$3^6 \cdot [a^+b^+c^+c^-b^-a^-; a^-c^+b^+]$
Parameterization TV14
Colouring (0 1 2) (0 1 2) (0 1 2)
Aspects 1, 2, 3
Rules
   $2 \leftarrow 1; 1$
   $3 \leftarrow 1; 6$
   $\vec{v_1} \leftarrow 2; 5$
   $\vec{v_2} \leftarrow 2; 3$

IH17

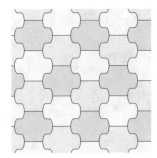

$3^6 \cdot [ab^+b^-ab^+b^-; ab^+]$
Parameterization TV10
Colouring (0) (1 2 0) (2 0 1)
Aspects 1
Rules
   $\vec{v_1} \leftarrow 1; 1$
   $\vec{v_2} \leftarrow 1; 2$

IH18

$3^6 \cdot [ababab; ba]$
Parameterization TV9
Colouring (0) (1 2 0) (2 0 1)
Aspects 1
Rules
   $\vec{v_1} \leftarrow 1; 1$
   $\vec{v_2} \leftarrow 1; 2$

IH20

$3^6 \cdot [aaaaaa; a]$
Parameterization TV9
Colouring (0) (1 2 0) (2 0 1)
Aspects 1
Rules
$\quad \vec{v_1} \leftarrow 1; 1$
$\quad \vec{v_2} \leftarrow 1; 2$

IH21

$3^4.6 \cdot [a^+b^+c^+d^+e^+; a^+c^+b^+e^+d^+]$
Parameterization TV15
Colouring (0 1 0 1 0 1) (1 2 0) (2 0 1)
Aspects 1, 2, 3, 4, 5, 6
Rules
$\quad 2 \leftarrow 1; 5$
$\quad 3 \leftarrow 2; 5$
$\quad 4 \leftarrow 3; 5$
$\quad 5 \leftarrow 4; 5$
$\quad 6 \leftarrow 5; 5$
$\quad \vec{v_1} \leftarrow 3; 2$
$\quad \vec{v_2} \leftarrow 4; 1$

IH22

$3^2.4^2 \cdot [a^+b^+c^+d^+e^+; b^-a^-e^+d^-c^+]$
Parameterization TV16
Colouring (0 1) (2 0 1) (0 1 2)
Aspects 1, 1r
Rules
$\quad 1r \leftarrow 1; 4$
$\quad \vec{v_1} \leftarrow 1r; 1$
$\quad \vec{v_2} \leftarrow 1r; 2$

IH23

$3^2.4^2 \cdot [a^+b^+c^+d^+e^+; a^+b^+e^+d^+c^+]$
Parameterization TV17
Colouring (0 1) (1 2 0) (2 0 1)
Aspects 1, 2
Rules
$\quad 2 \leftarrow 1; 1$
$\quad \vec{v}_1 \leftarrow 1; 3$
$\quad \vec{v}_2 \leftarrow 2; 4$

IH24

$3^2.4^2 \cdot [a^+b^+c^+d^+e^+; a^+b^+e^+d^-c^+]$
Parameterization TV17
Colouring (0 1 2 0) (2 0 1) (1 2 0)
Aspects 1, 2, 1r, 2r
Rules
$\quad 2 \leftarrow 1; 1$
$\quad 2r \leftarrow 2; 4$
$\quad 1r \leftarrow 2r; 1$
$\quad \vec{v}_1 \leftarrow 1; 5$
$\quad \vec{v}_2 \leftarrow 1r; 4$

IH25

$3^2.4^2 \cdot [a^+b^+c^+d^+e^+; b^-a^-e^+d^+c^+]$
Parameterization TV16
Colouring (0 1 2 0) (2 0 1) (0 1 2)
Aspects 1, 2, 1r, 2r
Rules
$\quad 2 \leftarrow 1; 4$
$\quad 2r \leftarrow 2; 2$
$\quad 1r \leftarrow 2r; 4$
$\quad \vec{v}_1 \leftarrow 1; 3$
$\quad \vec{v}_2 \leftarrow 1r; 2$

IH26

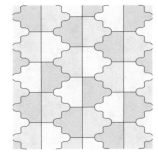

$3^2.4^2 \cdot [a^+a^-b^+cb^-; a^+b^-c]$
Parameterization TV18
Colouring (0 1) (1 2 0) (0 1 2)
Aspects 1, 2
Rules
$\quad 2 \leftarrow 1; 4$
$\quad \vec{v}_1 \leftarrow 1; 3$
$\quad \vec{v}_2 \leftarrow 2; 1$

IH27

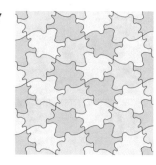

$3^2.4.3.4 \cdot [a^+b^+c^+d^+e^+; a^+d^-e^-b^-c^-]$
Parameterization TV19
Colouring (0 1 2 2) (0 1 2) (1 2 0)
Aspects 1, 2, 1r, 2r
Rules
$\quad 2 \leftarrow 1; 1$
$\quad 2r \leftarrow 2; 2$
$\quad 1r \leftarrow 1; 2$
$\quad \vec{v}_1 \leftarrow 2r; 5$
$\quad \vec{v}_2 \leftarrow 1r; 2$

IH28

$3^2.4.3.4 \cdot [a^+b^+c^+d^+e^+; a^+c^+b^+e^+d^+]$
Parameterization TV20
Colouring (0 1 2 0) (1 2 0) (0 1 2)
Aspects 1, 2, 3, 4
Rules
$\quad 2 \leftarrow 1; 1$
$\quad 3 \leftarrow 2; 3$
$\quad 4 \leftarrow 2; 4$
$\quad \vec{v}_1 \leftarrow 4; 4$
$\quad \vec{v}_2 \leftarrow 3; 3$

IH29

$3^2.4.3.4 \cdot [ab^+c^+c^-b^-; ac^+b^+]$
Parameterization TV21
Colouring (0 1 2 2) (1 2 0) (0 1 2)
Aspects 1, 2, 3, 4
Rules
$\quad 2 \leftarrow 1; 1$
$\quad 3 \leftarrow 1; 2$
$\quad 4 \leftarrow 1; 5$
$\quad \vec{v}_1 \leftarrow 3; 5$
$\quad \vec{v}_2 \leftarrow 4; 5$

IH30

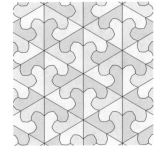

$3.4.6.4 \cdot [a^+b^+c^+d^+; d^+b^-c^-a^+]$
Parameterization TV22
Colouring (0 1 2 1 2 0) (1 2 0) (2 0 1)
Aspects 1, 2, 3, 1r, 2r, 3r
Rules
$\quad 2 \leftarrow 1; 1$
$\quad 3 \leftarrow 2; 1$
$\quad 1r \leftarrow 1; 2$
$\quad 2r \leftarrow 1r; 1$
$\quad 3r \leftarrow 2r; 1$
$\quad \vec{v_1} \leftarrow 3r; 3$
$\quad \vec{v_2} \leftarrow 3; 3, 2$

IH31

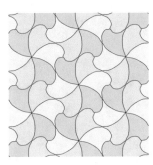

$3.4.6.4 \cdot [a^+b^+c^+d^+; d^+c^+b^+a^+]$
Parameterization TV23
Colouring (0 1 2 0 1 2) (0 1 2) (0 1 2)
Aspects 1, 2, 3, 4, 5, 6
Rules
$\quad 2 \leftarrow 1; 2$
$\quad 3 \leftarrow 2; 2$
$\quad 4 \leftarrow 3; 2$
$\quad 5 \leftarrow 4; 2$
$\quad 6 \leftarrow 5; 2$
$\quad \vec{v_1} \leftarrow 3; 1$
$\quad \vec{v_2} \leftarrow 5; 4$

IH32

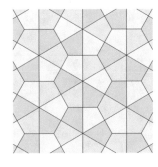

$3.4.6.4 \cdot [a^+b^+b^-a^-; a^-b^-]$
Parameterization TV23
Colouring (0 2 1 1 0 2) (1 2 0) (2 0 1)
Aspects 1, 2, 3, 4, 5, 6
Rules
$\quad 2 \leftarrow 1; 4$
$\quad 3 \leftarrow 2; 4$
$\quad 4 \leftarrow 1; 2$
$\quad 5 \leftarrow 4; 4$
$\quad 6 \leftarrow 5; 4$
$\quad \vec{v_1} \leftarrow 5; 3$
$\quad \vec{v_2} \leftarrow 2; 3, 3$

IH33

$3.6.3.6 \cdot [a^+b^+c^+d^+; d^+c^+b^+a^+]$
Parameterization TV24
Colouring (0 1 2) (0 1 2) (0 1 2)
Aspects 1, 2, 3
Rules
    $2 \leftarrow 1; 1$
    $3 \leftarrow 2; 1$
    $\vec{v}_1 \leftarrow 2; 2$
    $\vec{v}_2 \leftarrow 3; 3$

IH34

$3.6.3.6 \cdot [a^+b^+a^+b^+; b^+a^+]$
Parameterization TV24
Colouring (0 1 2) (0 1 2) (0 1 2)
Aspects 1, 2, 3
Rules
    $2 \leftarrow 1; 1$
    $3 \leftarrow 2; 1$
    $\vec{v}_1 \leftarrow 2; 4$
    $\vec{v}_2 \leftarrow 3; 1$

IH36

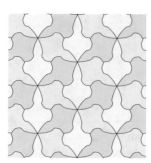

$3.6.3.6 \cdot [a^+a^-b^+b^-; b^-a^-]$
Parameterization TV24
Colouring (0 1 2) (0 1 2) (0 1 2)
Aspects 1, 2, 3
Rules
    $2 \leftarrow 1; 1$
    $3 \leftarrow 1; 4$
    $\vec{v}_1 \leftarrow 2; 1$
    $\vec{v}_2 \leftarrow 3; 3$

IH37

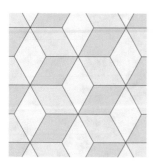

$3.6.3.6 \cdot [a^+a^-a^+a^-; a^-]$
Parameterization TV24
Colouring (0 1 2) (0 1 2) (0 1 2)
Aspects 1, 2, 3
Rules
    $2 \leftarrow 1; 1$
    $3 \leftarrow 2; 4$
    $\vec{v}_1 \leftarrow 2; 3$
    $\vec{v}_2 \leftarrow 3; 2$

IH38

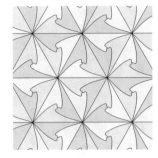

$3.12^2 \cdot [a^+b^+c^+; c^+b^-a^+]$
Parameterization TV25
Colouring (0 2 1 1 0 2) (1 2 0) (2 0 1)
Aspects 1, 2, 3, 1r, 2r, 3r
Rules
    $2 \leftarrow 1; 1$
    $3 \leftarrow 2; 1$
    $2r \leftarrow 2; 2$
    $3r \leftarrow 2r; 1$
    $1r \leftarrow 3r; 1$
    $\vec{v_1} \leftarrow 3r; 2$
    $\vec{v_2} \leftarrow 3; 2, 3, 2$

IH39

$3.12^2 \cdot [a^+b^+c^+; c^+b^+a^+]$
Parameterization TV25
Colouring (0 2 1 0 1 2) (2 0 1) (1 2 0)
Aspects 1, 2, 3, 4, 5, 6
Rules
    $2 \leftarrow 1; 1$
    $3 \leftarrow 2; 1$
    $4 \leftarrow 2; 2$
    $5 \leftarrow 4; 1$
    $6 \leftarrow 5; 1$
    $\vec{v_1} \leftarrow 3; 2, 1, 2$
    $\vec{v_2} \leftarrow 6; 2$

IH40

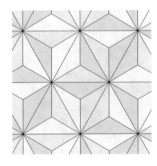

$3.12^2 \cdot [a^+ba^-; a^-b]$
Parameterization TV25
Colouring (1 0 2 2 0 1) (1 2 0) (2 0 1)
Aspects 1, 2, 3, 4, 5, 6
Rules
    $2 \leftarrow 1; 1$
    $3 \leftarrow 1; 3$
    $4 \leftarrow 1; 2$
    $5 \leftarrow 4; 1$
    $6 \leftarrow 4; 3$
    $\vec{v_1} \leftarrow 3; 2, 3, 2$
    $\vec{v_2} \leftarrow 2; 2, 3, 2$

IH41

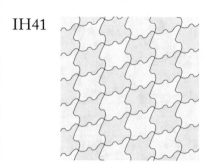

$4^4 \cdot [a^+b^+c^+d^+; c^+d^+a^+b^+]$
Parameterization TV26
Colouring (0) (1 0) (1 0)
Aspects 1
Rules
$\quad \vec{v_1} \leftarrow 1; 1$
$\quad \vec{v_2} \leftarrow 1; 2$

IH42

$4^4 \cdot [a^+b^+c^+d^+; c^+b^-a^+d^-]$
Parameterization TV26
Colouring (0 1) (1 0) (0 1)
Aspects 1, 1r
Rules
$\quad 1r \leftarrow 1; 2$
$\quad \vec{v_1} \leftarrow 1; 1$
$\quad \vec{v_2} \leftarrow 1r; 4$

IH43

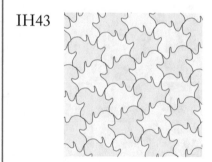

$4^4 \cdot [a^+b^+c^+d^+; c^-d^+a^-b^+]$
Parameterization TV26
Colouring (0 1) (1 0) (0 1)
Aspects 1, 1r
Rules
$\quad 1r \leftarrow 1; 1$
$\quad \vec{v_1} \leftarrow 1; 2$
$\quad \vec{v_2} \leftarrow 1r; 1$

IH44

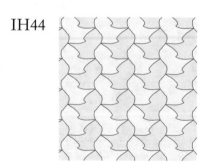

$4^4 \cdot [a^+b^+c^+d^+; b^-a^-d^-c^-]$
Parameterization TV27
Colouring (0 1) (0 1) (0 1)
Aspects 1, 1r
Rules
$\quad 1r \leftarrow 1; 1$
$\quad \vec{v_1} \leftarrow 1r; 1$
$\quad \vec{v_2} \leftarrow 1r; 3$

IH45

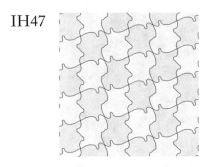

$4^4 \cdot [a^+b^+c^+d^+; c^-b^-a^-d^-]$
Parameterization TV28
Colouring (0 1) (0 1) (0 1)
Aspects 1, 1r
Rules
    $1r \leftarrow 1; 1$
    $\vec{v_1} \leftarrow 1r; 1$
    $\vec{v_2} \leftarrow 1r; 2$

IH46

$4^4 \cdot [a^+b^+c^+d^+; a^+b^+c^+d^+]$
Parameterization TV29
Colouring (0 1) (0 1) (0 1)
Aspects 1, 2
Rules
    $2 \leftarrow 1; 1$
    $\vec{v_1} \leftarrow 2; 3$
    $\vec{v_2} \leftarrow 2; 2$

IH47

$4^4 \cdot [a^+b^+c^+d^+; c^+b^+a^+d^+]$
Parameterization TV26
Colouring (0 1) (1 0) (0 1)
Aspects 1, 2
Rules
    $2 \leftarrow 1; 2$
    $\vec{v_1} \leftarrow 1; 1$
    $\vec{v_2} \leftarrow 2; 4$

IH49

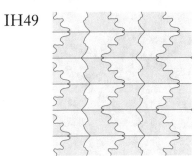

$4^4 \cdot [a^+b^+c^+d^+; a^-b^+c^-d^+]$
Parameterization TV31
Colouring (0 1 1 0) (0 1) (0 1)
Aspects 1, 2, 1r, 2r
Rules
    $2 \leftarrow 1; 2$
    $2r \leftarrow 2; 1$
    $1r \leftarrow 1; 1$
    $\vec{v_1} \leftarrow 1r; 3$
    $\vec{v_2} \leftarrow 2; 4$

IH50

$4^4 \cdot [a^+b^+c^+d^+; c^+b^-a^+d^+]$
Parameterization TV26
Colouring (0 1 1 0) (0 1) (1 0)
Aspects 1, 2, 1r, 2r
Rules
$\quad 2 \leftarrow 1; 4$
$\quad 2r \leftarrow 2; 2$
$\quad 1r \leftarrow 1; 2$
$\quad \vec{v}_1 \leftarrow 2r; 4, 2$
$\quad \vec{v}_2 \leftarrow 1; 1$

IH51

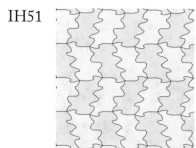

$4^4 \cdot [a^+b^+c^+d^+; c^-b^+a^-d^+]$
Parameterization TV32
Colouring (0 1 1 0) (0 1) (0 1)
Aspects 1, 2, 1r, 2r
Rules
$\quad 2 \leftarrow 1; 4$
$\quad 2r \leftarrow 2; 3$
$\quad 1r \leftarrow 1; 1$
$\quad \vec{v}_1 \leftarrow 1r; 1$
$\quad \vec{v}_2 \leftarrow 2; 2$

IH52

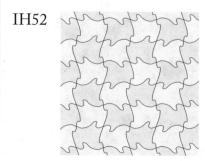

$4^4 \cdot [a^+b^+c^+d^+; c^-d^-a^-b^-]$
Parameterization TV30
Colouring (0 0 1 1) (0 1) (0 1)
Aspects 1, 2, 1r, 2r
Rules
$\quad 1r \leftarrow 1; 1$
$\quad 2 \leftarrow 1r; 4$
$\quad 2r \leftarrow 2; 1$
$\quad \vec{v}_1 \leftarrow 2r; 2$
$\quad \vec{v}_2 \leftarrow 1r; 1$

IH53

$4^4 \cdot [a^+b^+c^+d^+; b^-a^-c^+d^+]$
Parameterization TV33
Colouring (0 1 1 0) (0 1) (0 1)
Aspects 1, 2, 1r, 2r
Rules
$\quad$ 2 ← 1; 3
$\quad$ 2r ← 2; 1
$\quad$ 1r ← 1; 1
$\quad$ $\vec{v}_1$ ← 2; 4
$\quad$ $\vec{v}_2$ ← 1; 4, 1, 3, 2

IH54

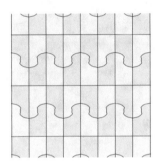

$4^4 \cdot [a^+b^+c^+d^+; a^-b^-c^-d^+]$
Parameterization TV34
Colouring (0 1 1 0) (0 1) (0 1)
Aspects 1, 2, 1r, 2r
Rules
$\quad$ 2 ← 1; 4
$\quad$ 2r ← 2; 3
$\quad$ 1r ← 1; 1
$\quad$ $\vec{v}_1$ ← 1r; 3
$\quad$ $\vec{v}_2$ ← 2; 2

IH55

$4^4 \cdot [a^+b^+c^+d^+; b^+a^+d^+c^+]$
Parameterization TV35
Colouring (0 1 0 1) (0 1) (0 1)
Aspects 1, 2, 3, 4
Rules
$\quad$ 2 ← 1; 2
$\quad$ 3 ← 2; 2
$\quad$ 4 ← 3; 2
$\quad$ $\vec{v}_1$ ← 2; 3
$\quad$ $\vec{v}_2$ ← 4; 4

IH56

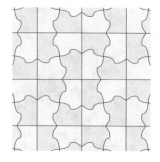

$4^4 \cdot [a^+b^+c^+d^+; b^+a^+c^-d^-]$
Parameterization TV36
Colouring (0 1 0 1 1 0 1 0) (0 1) (0 1)
Aspects 1, 2, 3, 4, 1r, 2r, 3r, 4r
Rules
$\quad$ 2 ← 1; 2
$\quad$ 3 ← 2; 2
$\quad$ 4 ← 3; 2
$\quad$ 2r ← 2; 3
$\quad$ 1r ← 2r; 1
$\quad$ 3r ← 2r; 2
$\quad$ 4r ← 3r; 2
$\quad$ $\vec{v_1}$ ← 1r; 4
$\quad$ $\vec{v_2}$ ← 2r; 4, 2

IH57

$4^4 \cdot [a^+b^+a^+b^+; a^+b^+]$
Parameterization TV26
Colouring (0) (1 0) (1 0)
Aspects 1
Rules
$\quad$ $\vec{v_1}$ ← 1; 1
$\quad$ $\vec{v_2}$ ← 1; 2

IH58

$4^4 \cdot [a^+b^+a^+b^+; a^-b^+]$
Parameterization TV26
Colouring (0 1) (1 0) (0 1)
Aspects 1, 1r
Rules
$\quad$ 1r ← 1; 1
$\quad$ $\vec{v_1}$ ← 1; 2
$\quad$ $\vec{v_2}$ ← 1r; 3

IH59

$4^4 \cdot [a^+b^+a^+b^+; b^-a^-]$
Parameterization TV37
Colouring (0 1) (0 1) (0 1)
Aspects 1, 1r
Rules
    $1r \leftarrow 1; 1$
    $\vec{v_1} \leftarrow 1r; 3$
    $\vec{v_2} \leftarrow 1r; 4$

IH61

$4^4 \cdot [a^+b^+a^+b^+; b^+a^+]$
Parameterization TV35
Colouring (0 1) (0 1) (0 1)
Aspects 1, 2
Rules
    $2 \leftarrow 1; 1$
    $\vec{v_1} \leftarrow 2; 1$
    $\vec{v_2} \leftarrow 2; 3$

IH62

$4^4 \cdot [a^+a^+a^+a^+; a^+]$
Parameterization TV35
Colouring (0) (1 0) (1 0)
Aspects 1
Rules
    $\vec{v_1} \leftarrow 1; 1$
    $\vec{v_2} \leftarrow 1; 2$

IH64

$4^4 \cdot [ab^+cb^-; cb^-a]$
Parameterization TV30
Colouring (0) (1 0) (1 0)
Aspects 1
Rules
    $\vec{v_1} \leftarrow 1; 1$
    $\vec{v_2} \leftarrow 1; 2$

IH66

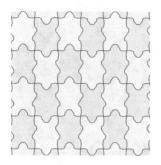

$4^4 \cdot [ab^+cb^-; cb^+a]$
Parameterization TV30
Colouring (0 1) (1 0) (0 1)
Aspects 1, 2
Rules
$\qquad 2 \leftarrow 1; 2$
$\qquad \vec{v_1} \leftarrow 1; 1$
$\qquad \vec{v_2} \leftarrow 2; 4$

IH67

$4^4 \cdot [a^+ba^-c; a^+bc]$
Parameterization TV28
Colouring (0 1) (0 1) (0 1)
Aspects 1, 2
Rules
$\qquad 2 \leftarrow 1; 1$
$\qquad \vec{v_1} \leftarrow 2; 2$
$\qquad \vec{v_2} \leftarrow 2; 4$

IH68

$4^4 \cdot [a^+b^+b^-a^-; b^-a^-]$
Parameterization TV37
Colouring (0) (1 0) (1 0)
Aspects 1
Rules
$\qquad \vec{v_1} \leftarrow 1; 1$
$\qquad \vec{v_2} \leftarrow 1; 2$

IH69

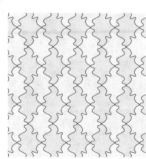

$4^4 \cdot [a^+a^-b^+b^-; a^+b^+]$
Parameterization TV27
Colouring (0 1) (0 1) (0 1)
Aspects 1, 2
Rules
$\qquad 2 \leftarrow 1; 1$
$\qquad \vec{v_1} \leftarrow 2; 2$
$\qquad \vec{v_2} \leftarrow 2; 4$

IH71

$4^4 \cdot [a^+b^+b^-a^-; b^+a^+]$
Parameterization TV35
Colouring (0 1 0 1) (0 1) (0 1)
Aspects 1, 2, 3, 4
Rules
$\quad 2 \leftarrow 1; 1$
$\quad 3 \leftarrow 2; 1$
$\quad 4 \leftarrow 3; 1$
$\quad \vec{v_1} \leftarrow 4; 3$
$\quad \vec{v_2} \leftarrow 2; 4$

IH72

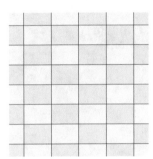

$4^4 \cdot [abab; ab]$
Parameterization TV30
Colouring (0) (1 0) (1 0)
Aspects 1
Rules
$\quad \vec{v_1} \leftarrow 1; 1$
$\quad \vec{v_2} \leftarrow 1; 2$

IH73

$4^4 \cdot [abab; ba]$
Parameterization TV35
Colouring (0 1) (0 1) (0 1)
Aspects 1, 2
Rules
$\quad 2 \leftarrow 1; 1$
$\quad \vec{v_1} \leftarrow 2; 1$
$\quad \vec{v_2} \leftarrow 2; 3$

IH74

$4^4 \cdot [a^+a^-a^+a^-; a^+]$
Parameterization TV37
Colouring (0) (1 0) (1 0)
Aspects 1
Rules
$\quad \vec{v_1} \leftarrow 1; 1$
$\quad \vec{v_2} \leftarrow 1; 2$

IH76

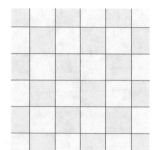

$4^4 \cdot [aaaa; a]$
Parameterization TV35
Colouring (0) (1 0) (1 0)
Aspects 1
Rules
$$\vec{v_1} \leftarrow 1; 1$$
$$\vec{v_2} \leftarrow 1; 2$$

IH77

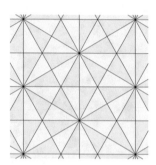

$4.6.12 \cdot [a^+b^+c^+; a^-b^-c^-]$
Parameterization TV38
Colouring (0 0 0 0 0 0 1 1 1 1 1 1) (0 1) (0 1)
Aspects 1, 2, 3, 4, 5, 6, 1r, 2r, 3r, 4r, 5r, 6r
Rules
$$1r \leftarrow 1; 3$$
$$2 \leftarrow 1r; 2$$
$$2r \leftarrow 2; 3$$
$$3 \leftarrow 2r; 2$$
$$3r \leftarrow 3; 3$$
$$4 \leftarrow 3r; 2$$
$$4r \leftarrow 4; 3$$
$$5 \leftarrow 4r; 2$$
$$5r \leftarrow 5; 3$$
$$6 \leftarrow 5r; 2$$
$$6r \leftarrow 6; 3$$
$$\vec{v_1} \leftarrow 4r; 1$$
$$\vec{v_2} \leftarrow 3; 1, 2$$

IH78

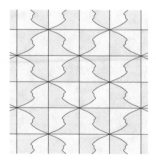

$4.8^2 \cdot [a^+b^+c^+; a^-b^+c^-]$
Parameterization TV39
Colouring (0 1 1 0) (0 1) (1 0)
Aspects 1, 2, 1r, 2r
Rules
$$2 \leftarrow 1; 2$$
$$2r \leftarrow 2; 3$$
$$1r \leftarrow 2r; 2$$
$$\vec{v_1} \leftarrow 1r; 3$$
$$\vec{v_2} \leftarrow 2r; 1$$

95

IH79

$4.8^2 \cdot [a^+b^+c^+; c^+b^+a^+]$
Parameterization TV41
Colouring (0 1 0 1) (1 0) (1 0)
Aspects 1, 2, 3, 4
Rules
    $2 \leftarrow 1; 1$
    $3 \leftarrow 2; 1$
    $4 \leftarrow 3; 1$
    $\vec{v}_1 \leftarrow 3; 2$
    $\vec{v}_2 \leftarrow 2; 2, 1$

IH81

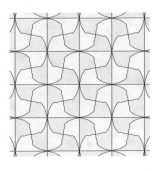

$4.8^2 \cdot [a^+b^+c^+; c^+b^-a^+]$
Parameterization TV41
Colouring (0 1 0 1 1 0 1 0) (0 1) (0 1)
Aspects 1, 2, 3, 4, 1r, 2r, 3r, 4r
Rules
    $2 \leftarrow 1; 1$
    $3 \leftarrow 2; 1$
    $4 \leftarrow 3; 1$
    $2r \leftarrow 2; 2$
    $3r \leftarrow 2r; 1$
    $4r \leftarrow 3r; 1$
    $1r \leftarrow 4r; 1$
    $\vec{v}_1 \leftarrow 3r; 2$
    $\vec{v}_2 \leftarrow 4r; 2, 1$

IH82

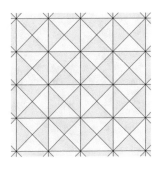

$4.8^2 \cdot [a^+ba^-; a^-b]$
Parameterization TV41
Colouring (0 1 0 1) (1 0) (1 0)
Aspects 1, 2, 3, 4
Rules
    $2 \leftarrow 1; 3$
    $3 \leftarrow 2; 3$
    $4 \leftarrow 3; 3$
    $\vec{v}_1 \leftarrow 3; 2$
    $\vec{v}_2 \leftarrow 2; 2, 3$

IH83

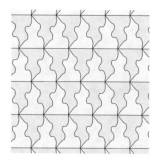

$6^3 \cdot [a^+b^+c^+; b^-a^-c^-]$
Parameterization TV42
Colouring (0 1) (0 1) (0 1)
Aspects 1, 1r
Rules
$\quad 1r \leftarrow 1; 3$
$\quad \vec{v}_1 \leftarrow 1r; 1$
$\quad \vec{v}_2 \leftarrow 1r; 2$

IH84

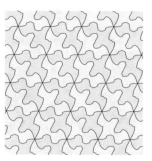

$6^3 \cdot [a^+b^+c^+; a^+b^+c^+]$
Parameterization TV40
Colouring (0 1) (0 1) (0 1)
Aspects 1, 2
Rules
$\quad 2 \leftarrow 1; 1$
$\quad \vec{v}_1 \leftarrow 2; 2$
$\quad \vec{v}_2 \leftarrow 2; 3$

IH85

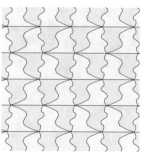

$6^3 \cdot [a^+b^+c^+; a^-b^+c^+]$
Parameterization TV40
Colouring (0 1 1 0) (0 1) (0 1)
Aspects 1, 2, 1r, 2r
Rules
$\quad 2 \leftarrow 1; 2$
$\quad 2r \leftarrow 2; 1$
$\quad 1r \leftarrow 2r; 2$
$\quad \vec{v}_1 \leftarrow 2; 3$
$\quad \vec{v}_2 \leftarrow 1r; 1$

IH86

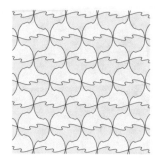

$6^3 \cdot [a^+b^+c^+; b^-a^-c^+]$
Parameterization TV42
Colouring (0 1 1 0) (0 1) (0 1)
Aspects 1, 2, 1r, 2r
Rules
$\quad 1r \leftarrow 1; 1$
$\quad 2r \leftarrow 1r; 3$
$\quad 2 \leftarrow 2r; 1$
$\quad \vec{v}_1 \leftarrow 1r; 1$
$\quad \vec{v}_2 \leftarrow 2; 3$

IH88

$6^3 \cdot [a^+b^+c^+; b^+a^+c^+]$
Parameterization TV43
Colouring (0 1 0 1 0 1) (0 1) (0 1)
Aspects 1, 2, 3, 4, 5, 6
Rules
    $2 \leftarrow 1; 1$
    $3 \leftarrow 2; 1$
    $4 \leftarrow 3; 1$
    $5 \leftarrow 4; 1$
    $6 \leftarrow 5; 1$
    $\vec{v_1} \leftarrow 4; 3$
    $\vec{v_2} \leftarrow 5; 3, 2$

IH90

$6^3 \cdot [a^+a^+a^+; a^+]$
Parameterization TV43
Colouring (0 1) (0 1) (0 1)
Aspects 1, 2
Rules
    $2 \leftarrow 1; 1$
    $\vec{v_1} \leftarrow 2; 2$
    $\vec{v_2} \leftarrow 2; 3$

IH91

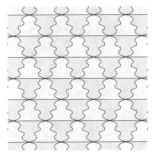

$6^3 \cdot [a^+a^-b; a^+b]$
Parameterization TV42
Colouring (0 1) (0 1) (0 1)
Aspects 1, 2
Rules
    $2 \leftarrow 1; 3$
    $\vec{v_1} \leftarrow 2; 1$
    $\vec{v_2} \leftarrow 2; 2$

IH93

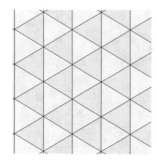

$6^3 \cdot [aaa; a]$
Parameterization TV43
Colouring (0 1) (0 1) (0 1)
Aspects 1, 2
Rules
$\quad 2 \leftarrow 1; 1$
$\quad \vec{v}_1 \leftarrow 2; 2$
$\quad \vec{v}_2 \leftarrow 2; 3$

Bibliography

[1] E. M. Arkin, L. P. Chew, D. P. Huttenlocher, K. Kedem, and J. S. B. Mitchell. An efficiently computable metric for comparing polygonal shapes. *IEEE Transactions on Pattern Analysis and Machine Intelligence*, 13:209–216, 1991. DOI: 10.1109/34.75509

[2] Robert Berger. The undecidability of the domino problem. *Memoirs of the American Mathematical Society*, 66, 1966.

[3] William W. Chow. Automatic generation of interlocking shapes. *Computer Graphics and Image Processing*, 9:333–353, 1979. DOI: 10.1016/0146-664X(79)90099-6

[4] Paul Church. Snakes in the plane. Master's thesis, School of Computer Science, University of Waterloo, 2008.

[5] Michael F. Cohen, Jonathan Shade, Stefan Hiller, and Oliver Deussen. Wang tiles for image and texture generation. *ACM Trans. Graph.*, 22(3):287–294, 2003. DOI: 10.1145/882262.882265

[6] John H. Conway, Heidi Burgiel, and Chaim Goodman-Strauss. *The Symmetries of Things*. A. K. Peters, 2008.

[7] Hallard T. Croft, Kenneth J. Falconer, and Richard K. Guy. *Unsolved Problems in Geometry*. Springer-Verlag, 1991.

[8] Karel Culik. An aperiodic set of 13 Wang tiles. *Discrete Math.*, 160(1-3):245–251, 1996. DOI: 10.1016/S0012-365X(96)00118-5

[9] L. Danzer, B. Grünbaum, and G. C. Shephard. Can all tiles of a tiling have five-fold symmetry? *American Mathematical Monthly*, 89:568–585, 1982. DOI: 10.2307/2320829

[10] Olaf Delgado Friedrichs. Data structures and algorithms for tilings i. 2003.

[11] Andreas W. M. Dress. The 37 combinatorial types of regular "Heaven and Hell" patterns in the euclidean plane. In H. S. M. Coxeter et al., editor, *M.C. Escher: Art and Science*, pages 35–45. Elsevier Science Publishers B.V., 1986.

[12] Douglas Dunham. Artistic patterns in hyperbolic geometry. In Reza Sarhangi, editor, *Bridges 1999 Proceedings*, pages 139–149, 1999.

[13] Douglas J. Dunham. Creating hyperbolic Escher patterns. In H. S. M. Coxeter et al., editor, *M.C. Escher: Art and Science*, pages 241–247. Elsevier Science Publishers B.V., 1986.

[14] M. C. Escher. *Escher on Escher: Exploring the Infinite*. Henry N. Abrams, Inc., 1989. Translated by Karin Ford.

[15] David W. Farmer. *Groups and Symmetry: A Guide to Discovering Mathematics*. American Mathematical Society, 1996.

[16] Natalie Priebe Frank. A primer of substitution tilings of the euclidean plane. *Expositiones Mathematicae*, 26(4):295–326, 2008. DOI: 10.1016/j.exmath.2008.02.001

[17] Chi-Wing Fu and Man-Kang Leung. Texture tiling on arbitrary topological surfaces. In *Proceedings of Eurographics Symposium on Rendering 2005 (EGSR 2005)*, pages 99–104, June 2005.

[18] Andrew Glassner. Andrew Glassner's notebook: Penrose tiling. *IEEE Computer Graphics & Applications*, 18(4), jul–aug 1998. ISSN 0272-1716. DOI: 10.1109/38.689670

[19] Solomon W. Golomb. *Polyominoes: Puzzles, Patterns, Problems and Packings*. Princeton University Press, second edition, 1994.

[20] Chaim Goodman-Strauss. Matching rules and substitution tilings. *Annals of Mathematics*, 147(1):181–223, January 1998. DOI: 10.2307/120988

[21] Chaim Goodman-Strauss. A small set of aperiodic tiles. *European Journal of Combinatorics*, 20:375–384, 1999. DOI: 10.1006/eujc.1998.0281

[22] Branko Grünbaum. The emperor's new clothes: Full regalia, G string, or nothing? *The Mathematical Intelligencer*, 6(4):47–53, 1984. DOI: 10.1007/BF03026738

[23] Branko Grünbaum and G. C. Shephard. Spherical tilings with transitivity properties. In Chandler Davis, Branko Grünbaum, and F. A. Sherk, editors, *The Geometric Vein: The Coxeter Festschrift*, pages 65–94. Springer-Verlag, New York, 1982.

[24] Branko Grünbaum and G. C. Shephard. *Tilings and Patterns*. W. H. Freeman, 1987.

[25] H. Heesch. Aufbau der ebene aus kongruenten bereichen. *Nachrichten von der Gesellschaft der Wissenschaften zu Göttingen*, pages 115–117, 1935. John Berglund provides an online English translation at http://www.angelfire.com/mn3/anisohedral/heesch35.html.

[26] H. Heesch and O. Kienzle. *Flachenschluss*. Springer–Verlag, 1963.

[27] Douglas Hofstadter. *Metamagical Themas: Questing for the Essence of Mind and Pattern*. Bantam Books, 1986.

[28] Daniel H. Huson. The generation and classification of tile-k-transitive tilings of the euclidean plane, the sphere, and the hyperbolic plane. *Geometriae Dedicata*, 47:269–296, 1993. DOI: 10.1007/BF01263661

[29] Owen Jones. *The Grammar of Ornament*. Studio Editions, 1986.

[30] Craig S. Kaplan. *Computer Graphics and Geometric Ornamental Design*. PhD thesis, Department of Computer Science & Engineering, University of Washington, 2002.

[31] Craig S. Kaplan. The trouble with five. *Plus Magazine*, December 2007.
http://plus.maths.org/issue45/features/kaplan/index.html.

[32] Craig S. Kaplan. Metamorphosis in Escher's art. In Reza Sarhangi, editor, *Bridges 2008 Proceedings*, pages 39–46, 2008.
http://www.cgl.uwaterloo.ca/~csk/papers/kaplan_bridges2008.pdf.

[33] Craig S. Kaplan and David H. Salesin. Escherization. In *Proceedings of the 27th annual conference on Computer graphics and interactive techniques (SIGGRAPH 2000)*, pages 499–510. ACM Press/Addison-Wesley Publishing Co., 2000. DOI: 10.1145/344779.345022

[34] Craig S. Kaplan and David H. Salesin. Dihedral Escherization. In *GI'04: Proceedings of the 2004 conference on Graphics interface*, pages 255–262. Canadian Human-Computer Communications Society, 2004.

[35] Johannes Kopf, Daniel Cohen-Or, Oliver Deussen, and Dani Lischinski. Recursive Wang tiles for real-time blue noise. *ACM Trans. Graph.*, 25(3):509–518, 2006. DOI: 10.1145/1141911.1141916

[36] Ares Lagae. *Wang Tiles in Computer Graphics*. Synthesis Lectures on Computer Graphics and Animation. Morgan & Claypool Publishers, San Rafael, CA, USA, March 2009. Editor: Brian A. Barsky, Series ISSN: 1933-8996 (print) 1933-9003 (electronic), Volume:4, Number: 1. DOI: 10.2200/S000179ED1V01Y200903CGR009

[37] Ares Lagae and Philip Dutré. An alternative for Wang tiles: Colored edges versus colored corners. *ACM Transactions on Graphics*, 25(4):1442–1459, October 2006. DOI: 10.1145/1183287.1183296

[38] Ares Lagae, Craig S. Kaplan, Chi-Wing Fu, Victor Ostromoukhov, Johannes Kopf, and Oliver Deussen. Tile-based methods for interactive applications. SIGGRAPH 2008 Class, SIGGRAPH 2008, Los Angeles, USA, August 2008. DOI: 10.1145/1401132.1401254

[39] Diego Nehab and Hugues Hoppe. Random-access rendering of general vector graphics. *ACM Trans. Graph.*, 27(5):1–10, 2008. DOI: 10.1145/1409060.1409088

[40] Victor Ostromoukhov. Mathematical tools for computer-generated ornamental patterns. In *Electronic Publishing, Artistic Imaging and Digital Typography*, number 1375 in Lecture Notes in Computer Science, pages 193–223. Springer-Verlag, 1998.

[41] Victor Ostromoukhov. Sampling with polyominoes. In *SIGGRAPH '07: ACM SIGGRAPH 2007 papers*, page 78. ACM, 2007. DOI: 10.1145/1276377.1276475

[42] Victor Ostromoukhov, Charles Donohue, and Pierre-Marc Jodoin. Fast hierarchical importance sampling with blue noise properties. In *SIGGRAPH '04: ACM SIGGRAPH 2004 Papers*, pages 488–495. ACM, 2004. DOI: 10.1145/1186562.1015750

[43] Zheng Qin, Michael D. McCool, and Craig S. Kaplan. Precise vector textures for real-time 3d rendering. In *I3D '08: Proceedings of the 2008 symposium on Interactive 3D graphics and games*, pages 199–206, New York, NY, USA, 2008. ACM. DOI: 10.1145/1342250.1342281

[44] Doris Schattschneider. Escher's classification system for his colored periodic drawings. In H. S. M. Coxeter et al., editor, *M.C. Escher: Art and Science*, pages 82–96. Elsevier Science Publishers B.V., 1986.

[45] Doris Schattschneider. *M.C. Escher: Visions of Symmetry*. Harry N. Abrams, Inc., second edition, 2004.

[46] Marjorie Senechal. *Quasicrystals and Geometry*. Cambridge University Press, 1996.

[47] G. C. Shephard. What Escher might have done. In H. S. M. Coxeter et al., editor, *M.C. Escher: Art and Science*, pages 111–122. Elsevier Science Publishers B.V., 1986.

[48] A. V. Shubnikov and V. A. Koptsik. *Symmetry in Science and Art*. Plenum Press, 1974.

[49] Jos Stam. Aperiodic texture mapping. Technical Report 01/97-R046, The Europeran Research Consotium for Informatics and Mathematics, 1997.

[50] Martin von Gagern and Jürgen Richter-Gebert. Hyperbolization of euclidean ornaments. *The Electronic Journal of Combinatorics*, 16(2), 2009–2010.

[51] Dorothy K. Washburn and Donald W. Crowe. *Symmetries of Culture*. University of Washington Press, 1992.

[52] Hermann Weyl. *Symmetry*. Princeton Science Library, 1989.

[53] Jane Yen and Carlo Séquin. Escher sphere construction kit. In *Proceedings of the 2001 symposium on Interactive 3D graphics*, pages 95–98. ACM Press, 2001. DOI: 10.1145/364338.364371

Biography

CRAIG S. KAPLAN

Craig S. Kaplan is an associate professor in the David R. Cheriton School of Computer Science at the University of Waterloo. He is interested in applications of computer graphics in art, architecture and design, and makes occasional forays into human-computer interaction and computational geometry. Craig has a Bachelor's degree in Pure Mathematics and Computer Science from the University of Waterloo and a PhD in Computer Science from the University of Washington. A native of Montreal, he lives in Waterloo, Ontario with his long-suffering wife, their daughter and son, and two cats. He enjoys solving puzzles, making paper airplanes for his children, and eating soup.

Printed in the United States
by Baker & Taylor Publisher Services